Clarence Hawkes:
America's Blind Naturalist and the World He Lived In

CLARENCE HAWKES:

America's Blind Naturalist and the World He Lived In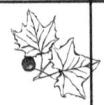

JAMES A. FREEMAN

Genealogy House
Amherst, Massachusetts

© 2009 James A. Freeman. All rights reserved.

No portion of this book may be reproduced or used in any form, or by any means, without prior written permission of the publisher.

First published October 24, 2009
Proclaimed Clarence Hawkes Day throughout the Commonwealth of Massachusetts by Governor Deval Patrick

Genealogy House
P.O. Box 3561
Amherst, MA 01004
www.genealogyhouse.net
Cover and interior designed by Vicki Stiefel (*www.lightplayphotos.com*)

Printed in the United States of America

ISBN: 978-1-935052-21-0

A condensed version of this material first appeared in *Cultivating a Past: Essays on the History of Hadley, Massachusetts*, ed. Marla R. Miller. Amherst, MA: University of Massachusetts Press, 2009.

Library of Congress Cataloging-in-Publication Data

Freeman, James A., 1935-

Clarence Hawkes : America's blind naturalist and the world he lived in / by James A. Freeman.

p. cm.

Includes bibliographical references.

ISBN 978-1-935052-21-0 (pbk.)

1. Hawkes, Clarence, 1869-1954. 2. Hawkes, Clarence, 1869-1954--Influence. 3. Naturalists--United States--Biography. 4. Authors, American--Biography. 5. Blind--United States--Biography. 6. Natural history literature--United States--History. 7. American literature--19th century--History and criticism. 8. American literature--20th century--History and criticism. 9. Hadley (Mass.)--Biography. 10. Hadley (Mass.)--History. I. Title.

CT275.H4735F74 2009

974.4'23043092--dc22

[B]

2009041435

Once again, to Margaret

But also to the future of Eric and Maya, Nico and Kai

Contents

- 11 Preface
- 15 The Life of Clarence Hawkes
- 51 The Writings:
  - Poetry
  - Nature Works
  - Novels
- 81 Hawkes' Reputation
- 85 How Hadley Now Remembers Hawkes
- 95 The Lesson of Clarence Hawkes
- 97 Clarence Hawkes Genealogy
- 101 Appendix: Nature Writing Before Hawkes
- 107 Clarence Hawkes Bibliography
- 112 Notes
- 113 Endnotes

"Unfortunate the boy or girl who grows up, or has grown up, without reading about Shovelhorns, the Moose monarch; Shaggycoat, the astute beaver; Black Bruin, the genial bear, and a score of other wild personages whose biographies have been set down by the typewriter of painstaking Clarence Hawkes."
Time 7.16 (19 April 1926): 40

Preface

On Saturday, 24 October 2009, Governor Deval Patrick proclaimed Clarence Hawkes Day throughout the Commonwealth of Massachusetts. The ceremony on Hadley's famous mile-long common evoked memories of the remarkable blind author for many in the audience. Some mature listeners had helped Hawkes proofread his poems, nature books, memoirs and novels before his death in 1954. Others remembered the international fame that nearly 60 books had earned for himself and towns most closely associated with him: Goshen, where he was born in 1869, Ashfield, and, most especially, Hadley. Still others in the audience, born after the height of Hawkes' reputation, wanted to know more about his triumphs over adversity. How few of us could survive rural poverty, losing half our left leg at nine and being blinded at thirteen? Although related to the eminent poet William Cullen Bryant and to the first sitting President of what would soon become the University of Massachusetts, William Clark Smith, he had to make his own way. He aroused affection and high regard from college presidents, Helen Keller, Calvin Coolidge, naturalists and uncounted millions of readers. Like his near contemporary Laura Ingalls Wilder, author of the Little House on the Prairie books, he transmuted pain into art. Hawkes' never-failing optimism did not ignore loss in the human or animal world, but he made it a necessary partner to an encompassing vision.

Such nobility under pressure wins admiration in all eras; artistic reputation, however, often fades. Through no fault of his own, Hawkes' prominence, like that of Henry Wadsworth Longfellow and James Whitcomb Riley, has receded from popular memory. The twentieth century elevated cryptic poetry over easily understood verse; dynamic, multi-sensory, wide-screen depictions of nature over straightforward narratives; agonized confessions in autobiography over a life presented as purposeful, and experimental prose over Victorian novels that had beginnings, middles and ends. What readers in the twenty-first century can recover from Hawkes' writings is a model to emulate. His goal of integrating the human and the natural worlds seems even more important today as the globe warms and species vanish. His gift of transmitting information about nature called forth the praise cited above. Hawkes' similarly enthusiastic contributions to civic life certainly have earned him the honor of being the focus of attention for an entire state. May he give others the courage to live together in unselfish harmony.

The Life of Clarence Hawkes
(1869-1954)

In 1943 United Press supplied its 1400 papers with an article honoring Clarence Hawkes (1869-1954), proof that the blind Hadley author had touched the lives of uncounted people throughout the world. He represented in real life the kind of person that he praised in more than three score books. Born on a marginal farm, with little choice for a vocation other than subsistence agriculture, he overcame two tragic boyhood accidents—losing part of his left leg at nine and his sight at thirteen—to become one of America's outstanding writers of nature books. In addition, his poems, novels, and autobiographies were translated into Braille, French, Finnish, German, Danish, Chinese, and Japanese, unusual honors for any author. He likewise gained respect for his abilities as a public speaker, his knowledge of sports—especially fishing, baseball, and football—and his participation in political events.

Hawkes accomplished much in a time when medicine and social welfare systems often failed the injured, dooming them to lonely misery. During his long life, robber barons and powerful trusts exploited the helpless, both native and immigrant; two world wars again demanded unthinkable sacrifices from a country that had already suffered a cruel civil war; a global depression destroyed countless families; and heedless hunting threatened many species with extinction.

Despite his personal tragedies and the tumultuous times, Hawkes remained positive about the large world. He had to write to survive, but he knew best selling books about nature presented falsehoods that misinformed readers. Still, he trusted that honesty could withstand the unfair attacks of fate and of "nature fakery." His personal qualities and his books urged millions of readers around the globe to connect all life—animal, bird, and plant—with their own. Bonding to the natural world would in turn eliminate the mysteries of a discontinuous planet. Once enlightened, we could link with our Creator, promote social compassion, and achieve personal integrity. Hawkes' vision of a cohesive physical and social universe duplicated that of American Transcendentalists as well as that of far-away Mohandas Gandhi, also born in 1869.

His career copied the optimistic pattern of Horatio Alger's poor boys who resisted the perils of urban life to finally become wealthy. Hawkes' triumphs, however, were rural, artistic, and moral, not financial. His ideal reader learned that true victory should be defined as a life governed, not by Ragged Dick's striving for personal wealth, but (as he often put it) "Patience, Perseverance, and Pluck," traits that merged private and social needs with the animate world.

Realistic but never bitter, Hawkes displayed inexhaustible energy. He wore out nine or ten typewriters (who could keep track during his busy career?). Like the tireless reindeer in his 1915 *King of the Flying Sledge*, who ran 300 miles in 24 hours during a bitter Norwegian winter to bring a doctor to a sick child, he researched, remembered, and wrote throughout his career with awesome diligence, always showing that no creature is isolated from a community. Widely known as the lame, blind author, he wanted to be accepted as a member of a family: "please do not think of me, when you read, as a crippled, groping individual; ... but think of me as a brother and a friend, ... bubbling over with a sense of humor and scattering sunshine always."[1]

Although conventionally religious, deeply patriotic, and conservative politically, he did not scorn those who differed from him. During an era of easy bigotry against newer citizens, he wrote a moving elegy for the death of a young Polish neighbor boy crushed under a tobacco rack,[2] a short story sympathizing with a winsome Polish girl in Hadley who had to return to her birthplace because of prejudiced immigration laws,[3] and a fond account of romance between Italian immigrants.[4] His poems avoided self-congratulation as they spoke out in straightforward language for the underprivileged. The novels, too, explored ways that hard hearts might be redeemed.

Hawkes' special qualities encouraged everyone he touched. Helen Keller valued his friendship, Teddy Roosevelt and fellow naturalists admired his knowledge of animals, college presidents and state governors praised his achievements, Calvin Coolidge kept Hawkes' picture on his desk starting in 1895,[5] readers, ordinary or blind, American or foreign, thanked him because he inspired them, and Hadley neighbors appreciated his contributions to town life. His widely reviewed books were favorably compared with works by Kipling and advertised along with volumes by Robert Frost. In 1937 one commentator estimated that he had "won an audience of five million young people around the world."[6] Six years later *The Boston Sunday Post* estimated that his "newspaper and magazine articles, plus an occasional talk via radio," had reached at least one million more people.[7] His 60-some volumes may no longer be on family bookshelves, but they confirm his life-long commitment to describe truthful encounters between people and the animate world.

The following pages discuss Hawkes' life, his works—poetry, animal tales, and novels—his reputation, and the memories of current Hadley residents. For the reader who wishes expanded information about specific topics, an Appendix on Nature Writing Before Hawkes, genealogical charts, a bibliography, and endnotes follow these narrative and evaluative sections.

The Life of Clarence Hawkes

A pessimist might have thought that Hawkes' effective life would end at age 13. Born during a blizzard in a western section of Goshen, Massachusetts, on 16 December 1869, he reflected, "I have always been sorry that I was not born in June instead of December. In its serenity and beauty, instead of the storm and hardship of winter."[8] However, he never held his unpromising entry into the world responsible for later setbacks. Rather, he made light of its harshness by calling to mind

> a humble little cottage in Lithia, then called Road Island. This was not a very auspicious domicile in which to be born, but it was the best I could do under the circumstances. I was in something of a hurry at the time and did not have a chance to pick and choose, in the vernacular of airmen[,] I made what might be called a forced landing.[9]

The lowly house was all that his father Enos Smith Hawkes could afford. Today it exists only as a cellar hole, so familiar to many New England landscapes. In the early 1940s Hawkes returned and noted, "Much of the pasture land I knew as a boy, is now grown up to brush or small timber."[10] Today tall trees have grown up to block the view young Clarence would have had of the deep valleys and distant hills in the area of Goshen, Ashfield, and Cummington, vistas that enriched his early days.[11]

The boy needed such benefits because father Enos (1840–1894) had few financial or physical resources. The son of William Hawkes and Almira Smith, he was descended from John Hawks, one of Hadley's founders, and related to William Smith Clark, first sitting president of the University of Massachusetts, as well as to William Cullen Bryant, the eminent poet of Cummington. Enos, however, inherited little. He and his brother William Smith Hawkes had gone west in the 1850s, a common destination for aspiring young people, but he obviously did not prosper. They enlisted in the Union army from Coloma, Whiteside County, Illinois, on 7 September 1861. Enos served less than a year as a private in Company A, 34th Illinois Volunteer Infantry until he was "discharged for disability 19 August 1862 at Battle Creek Tenn for disease of eyes."[12]

His time as "one of Uncle Sam's boys in blue in '61" was disastrous because (as his son put it, somewhat evasively) "hardtack had spoiled his digestion."[13] Enos' experience left him (again in his son's words) "a broken-down Civil War veteran."[14] He returned and occupied the house in Goshen on Loomis Road owned by relations, the Willcutt family, but commonly called the Allen house. The government may have helped him pay the rent: on 20 March 1864 he began to receive a pension of $2.00 per month for "amaurosis," the occasional blindness that oddly prefigured his son's ailment.[15] Apparently the whole pension was not paid until nearly two decades later when Enos was again placed on the roll, this time "at $24 per month, and has recently received $3525.00 as arrearages."[16]

His stay in Goshen from the early 1860s until 1879 made little impact on the tiny town, although an 1873 map clearly denominated the "E. Hawks" residence.[17] Many other Hawkes lived in the vicinity, yet histories do not spotlight him.[18] A compromised man, Enos "made the training of dogs and hunting almost his entire business for years."[19] He earned gratitude from his young son for teaching him "of the great out-of-doors world and the ways of all the furred and feathered folks of the New England Woods."[20] Hawkes' writings may covertly recall his father when they honor the courage and discipline of soldiers while at the same time lamenting war's waste and the over hunting of wildlife. For some reason, Hawkes does not mention that his father had secured three patents, one for a game protector, a second for a painting machine, and a third for a scrubbing machine.[21]

Hawkes' mother, Edlah Betsey Olive Bates Gurney (1849—1899), had, like Enos, been born in Ashfield, Franklin County. They married in Ashfield on 29 November 1866 and spent young Clarence's early years on Enos' "little rock-ribbed, sterile farm in the western part of the town." The census of 1860 listed 439 inhabitants, while that of 1870 counted only 368, a hint that the region lacked possibilities.[22] The boy's isolation was enlivened by visits to his beloved maternal grandmother, Mrs. Josiah Gurney (Emily Bates Jenkins. c. 1820—1877), in Ashfield, "an adjoining town made famous by Charles Eliot Norton and George William Curtis, and their Sanderson academy dinners." She awakened his appreciation "for birds and squirrels."[23] The opportunity to hear enlightening speakers sparked his early career as a lecturer.

Even though young Clarence enthusiastically observed the countryside around the farms with his intense blue eyes, he differed little from his younger sister, Alice Edith (1872—1897), or three brothers, Enos Raymond (1875—?), Arthur Josiah (1877—1941), and later born Ernest William ("Kid." 1881—1957). Soon fate made him special. Returning from Spruce Corner School in June 1879, when only nine, he jumped off a stone wall near their house, injuring his left foot. After weeks of infection, doctors came to the house on 29 July. They began to operate before the anesthetic took effect and he saw his own blood spurt onto the wallpaper; when he woke, he found his leg amputated below the knee. The family moved in September from Goshen to the Gurney farm in Ashfield where they lived with their widowed grandfather Josiah (1806—1886). Horatio and Emily Willcutt Culver took over the Goshen place, so Hawkes literally and figuratively could not go home again.

Showing the determination that distinguished him, he managed to walk, first on crutches, then on a "peg leg," and tried to recapture a normal life. He boated, drove teams, snow shoed, pitched in baseball games, and even ice skated. In his books, both humans and beasts suffer injuries, but survive as he had. Hal Houghton, the teenage Montana lad who owns a firefighting horse, badly sprains his ankle yet manages to mount and ride back to the ranch so it can mend.[24] Hal's animal counterpart, the arctic fox, might have to "gnaw off its own leg after being caught [in a spring trap] and limp away to freedom on three legs. It would always be a cripple among its kind, but that was often deemed better than death beneath the club of a trapper."[25] Such recoveries sounded reasonable to a stoic generation that knew Teddy Roosevelt, having broken his arm while on a fox hunt in 1895, even so remounted his horse, ate a hearty dinner, and hiked next day in the woods for three hours.[26] For a brief four years, young Hawkes seemed to have regained some autonomy in his community.

Then fate once more assaulted him. While hunting with his father near Spruce Corner, Ashfield, about noon on Tuesday, 12 August 1883, young Hawkes rested some fifteen rods away amid bushes, invisible to his father. When a woodcock swooped downward, Enos fired, lacerating the thirteen year old's face with "mustard seed" shot. As Hawkes put it, "I looked for the last time with full vision upon Mother Earth and all her visible beauties."[27]

No matter what guided the father's hand—pure accident, lunacy, vertigo, stress caused by the assault of Southern chauvinists, or money worries—the son's immediate pain, shock, bleeding, faintness, and vomiting began two years of suffering. For six weeks after the mishap, he waited for his two broken fingers to heal and depended upon ice packs to relieve the misery in his eyes. Sadly, he sensed that sight was fading: after only two weeks, a window at the foot of a stairs no longer lit them and he had to count in order to descend.

Again victimized, the feisty lad now faced lifelong blindness. Later, his fiction would imagine rescue scenes that turned away such disasters. The closest apotropaic incident, in *Big Brother*, recreates the calamitous August and the forest of Ashfield when Tim the circus boy, also nearly fourteen years old, escapes to woods with his pet bear. The two fugitives, "resting by a fallen log," are overtaken by "a young man with a gun in his hand": "Tim was too horrified to cry out and the hunter did not see the boy but just the mighty form of the bear. With a movement so quick Tim could scarcely follow it, he snapped the gun to his shoulder and fired. A charge of bird shot sent half a dozen leaves floating down on Tim's head but did no further damage." Big Brother snarls: "This was enough for the hunter and with a yell of terror he threw his gun into the bushes and fled as though the bear himself was after him."[28] But such an alternative scenario could not undo the real life blinding.

Doctors at home and in Boston tried to restore his sight. The method now seems barbaric, yet at the time inserting needles into the injured eyeballs with no anesthetic was felt to be a possible cure. Interviewed four decades later by Bruce Barton, soon to be one of the giants of American advertising, Hawkes vividly recreated the horror of this odd regimen:

> I do not like to dwell on the memory of the next two years.... There could be no anesthetic, for the object of the operation was to torture the eye so that it might, in a sort of frenzied self-defense, put forth new powers.... They strapped me to a table, put a rubber blanket under my head and tied my hands. Then there was brought into play a fiendish little machine which gripped the eye at every point where the muscles control it.... Imagine then a sharp lancet thrust slowly into the very center of the eyeball, down and down, until it must have reached the center of the brain.[29]

When the physicians finally admitted that there was no hope, Hawkes felt almost relieved: at least the pain would stop. His mother did what she could to assuage his loss, reading to him from classic authors, tending his needs, and imprinting her devotion on his tenacious memory.

To a typical onlooker in the 1880s, Hawkes' first one-and-one-half decades had already provided all the independence he could expect. The newspaper report of his accident summed up the conventional response to a handicapped male in an agricultural setting: sightlessness "makes him certainly an object of commiseration."[30] In fiction and real life, Victorians reacted to disabilities with three emotions: pity (think of Charles Dickens' lame Tiny Tim in *A Christmas Carol*, 1843); sadness (the blind flower seller Nydia in Edward Bulwer-Lytton's *The Last Days of Pompeii*, 1834); or horror (Blind Pew the pirate, trampled to death by horses, and rascally Long John Silver, the one-legged pirate, both in Robert Louis Stevenson's *Treasure Island*, 1883).[31]

The plight of his own father had daily taught him the poignant effects of an injury. When Hawkes lost his leg and his sight, no financial aid appeared. The family had already "spent about a thousand dollars for doctors" and had few resources for future care.[32] Even famous Helen Keller and her teacher, Annie M. Sullivan, depended on the kindness of strangers who launched an appeal for funds to place the two "in a position of permanent financial independence."[33] By any sensible standard, the insignificant boy from the country had no life ahead of him. As one reviewer said thirty-three years after the shooting, "For most of us, the coming of blindness would mean the fall of the curtain, the end of the drama."[34]

Yet Hawkes' career exploded negative clichés. Summoning his typical resolve to conquer despair, he enrolled, although hesitantly, in Watertown's Perkins Institute for the Blind in fall 1885. Such a move took courage because, as Charles Winchell from Dalton, Massachusetts, who entered the school a decade later in 1896, observed, "very few people in western Massachusetts in those days knew much about Perkins."[35] Fortunately, the state paid his tuition.

Perkins Institute, South Boston.

PERKINS INSTITUTE

Even more fortunately, Hawkes, once the ultimate outdoor lad, thrived under the indoor routine. A headline writer for the *Boston Sunday Post* aptly called Perkins the "House of Joyous Darkness" for him.[36] His very first triumph—feeling the outline of a wooden puzzle piece that he recognized as Cuba—foretold his later fascination with far-away places and, perhaps, his support of the Spanish-American War. According to a report card that looked back on the academic year ending 25 June 1889, his lowest grade for Physics, Geometry, Arithmetic, History, and Language was 9 on a scale of 10. An encouraging note from the Acting Director said that he "Ought to go to college."[37] He may have edited *The Echo*, Perkins' newspaper, and printed his first poem in it.

Graduating on 3 June 1890, Hawkes gave a talk, "The

Future of the Colored Man." It reflected the current hope of Abolitionists that the Civil War had allowed both races to advance together. In addition, it predicted his future sensitivity to the concerns of his audiences. Even a quarter century after the conflict, the dilemma of Blacks haunted northerners. Graduation speeches at Hopkins Academy, Hadley's high school, echoed these concerns: in 1890 Edward C. Pelissier spoke on "The Negro Problem"; in 1894 James Robert Ryan spoke on "Education of the Negro."[38] Hawkes may have been present when the popular speaker Henry Woodfin Grady (1850—1889), editor of *The Atlanta Constitution*, delivered "The Race Question" at the Vendome to the Boston Merchants' Association on 12 December

HAWKES' NOCTOGRAPH

1889. The address was widely reprinted and soothed the consciences of wary northerners by citing the current social advantages enjoyed by people of color down South.[39] The root idea— "patience…confidence…sympathy …loyalty" (as Grady put it)—may help even damaged outsiders to triumph certainly heartened the rural lad who later adopted "Patience, Perseverance, and Pluck" as his motto. Hawkes' oration, which quoted liberally from Grady, elicited praise from several Boston newspapers: it "showed evidence of study and intelligent thought on the part of the young man, and was all the more effective for his infirmity"[40]; it was "a thoughtful and suggestive essay."[41]

Perkins made possible Hawkes' long-standing friendship with Helen Keller, who was in the class behind him. The education was so liberating that he stayed in the city for further lessons in music and oratory and, at the Emerson School, elocution, plus some lectures on law, apparently at Boston University. His memory of the new joy in an expanded cosmos may reappear symbolically in his subsequent account of wild blue fox cubs in the arctic.

> For nine days they were totally blind, then their eyes came slowly open, and they could see the faint light in the burrow, … Finally, when they were two months old the mother fox took them outside into the great wide world. When they saw for the first time that the narrow, stuffy burrow was not all there was of the universe they were much surprised. They winked and blinked at the bright light.[42]

A debilitating attack of grippe in spring 1890 cut short the post-graduate program. At 21 he moved back to his family, which had resided in Cummington since the Ashfield farm was sold in spring 1885. The house that his father bought, built shortly before 1873, still stands on Main Street. Although Hawkes could not see them, the village possessed a handsome church, useful library, nicely tended cottages, and pleasant views of the surrounding hills.[43] A photograph of himself that he donated to Cummington's Historical Commission has a fond inscription on the back: "Presented to the Bryant Library where I got my first inspiration to write."

He needed all the resources he could muster when he began the tough job of supporting himself. Hawkes reasoned that his neighbors might enjoy respite from the harsh routine of backcountry life while attending lectures like those available at Norton's Sanderson Academy in Ashfield and at Chautauqua, New York. He could speak about American literature. According to his own recollections, the years from 1891-1895, "knocking about lecturing,"[44] required enormous effort, physical and emotional. He had to brave rain and white-out sleet storms, small audiences and meager pay, yet he lived up to his commitments. One fatiguing address earned him $1.65, though the reviews of his talks predicted a lifetime of admiration. His address on "the chief American poets" at Ashfield's Congregational chapel "was a fine literary treat... finely rendered." As usual, the performance had a moral purpose. He accepted the normal linkage of church venues, "high" culture, entertainment, and ethical improvement. Hawkes "closed with an exhortation to the young to do with their might whatever they had to do." Verbally, he may have been echoing the famous encouragement of William Smith Clark to his students in Hokkaido, Japan: "Boys, be ambitious."

Inspired by perennial giants such as Dante and Shakespeare as well as popular James Whitcomb Riley,

Henry Wadsworth Longfellow, and his Cummington predecessor/relative William Cullen Bryant, Hawkes readied his own verse for publication. Despite numerous rejections, he continued "bombarding newspapers and magazines with poems" so that "in 1895 it was said that I had more poems published in leading weeklies and magazines than any other poet in America."

HAWKES' HOME

To publish a book, however, he had to guarantee 300 subscribers. With characteristic courage, during a blisteringly hot August, he hired five boys to drive to far-flung dwellings in the Hampshire hills and collect enough pledges so that the first of five collections, *Pebbles and Shells*, could appear in 1895.Like most of Hawkes' poetry, it employed the predictable metrical forms, sentimental content, and plain diction familiar to Victorian ears. The volume earned a reputation and so he could more easily follow it with four others: *Three Little Folks: Verses for Children* (1896), *Idyls of Old New England* (1897), *Songs for Columbia's Heroes: War Poems for 1898* (1898), and *The Hope of the World* (two editions, 1900). He kept his

HAWKES' HOUSE TODAY

name before the public in subsequent years by printing separate poems in newspapers and magazines, most notably a series of annual Christmas greetings during the early 1930s in papers from Boston, Springfield, Holyoke, and Northampton. Many appeared in the 1938 collection *Christmas All the Year* and the 1939 *Holiday Hopes*. These compositions, totaling perhaps 2000,[47] found their way into anthologies and earned him his usual epithet, "the blind poet of Hadley."

During spring 1892, the family moved from Cummington to Hadley, where he spent the remainder of his productive life. He obviously felt he had reached a nurturing haven. Ancestor John Hawks had helped found the town in 1659. His beloved sister Alice graduated from Hopkins in the class of 1895, having sung soprano in the Choral Club and won a prize for elocution, little suspecting that she would die two short years later.[48] A two-story house on the east side of the scenic common (number 50 West Street then, now 104) would shelter him for the rest of his life. That residence, relocated in the 1870s to accommodate the new route of the Massachusetts Central Railroad, formerly belonged to Reverend E. S. Dwight, the last pastor of the Russell Church. "My home," Hawkes fondly explained, "faces upon the broadest and most beautiful street in the world, which is flanked by four rows of enormous elms."[49] Its location, on a mile long common between a loop in the Connecticut River, allowed him and a friend to launch a skiff from the upstream landing (now known as the Dike) and enjoy a leisurely float, fishing or sensing "the most beautiful seven miles of river scenery" that included, in his day as in ours, "green, fertile meadows, with the twin mountains of Holyoke and Tom dreaming in the distance."[50]

Although finances were still limited, he married Bessie Wilder Bell (1869—1958) at 9 a.m. on Monday, 30 October 1899. She was the artistic daughter of Samuel Reuben Bell (1839—1920), a tobacco dealer, and Sarah A. Wilder Bell (c.1843—1913).[51] Bessie probably studied at Hopkins

Academy (although I have not located her name in any of the perhaps incomplete histories of the school). Brooklyn's famed Pratt Institute admitted her on 25 September 1893. According to blurry microfilm records, she took courses in "Instrumental Perspective" and "Drawing—figure from life."[52] Also, she studied with Bruce Crane, a popular landscape artist living in New York City who specialized in bucolic scenes empty of people, but glowing with autumnal hues.

The wedding took place in her parents' home at 51 (now 107) West Street, a few yards south of state route 9, directly opposite Hawkes' own house.[53] At that time, as he wryly put it, "my bank account was again a minus quantity."[54] The newspaper account of the wedding hinted at the straightened circumstances of the young couple: although "Both are well known and popular young people," only the immediate families attended.[55] Hawkes' mother, ailing for at least four years, had recently died, so to pay for her medical expenses and funeral, he had been obliged to give lectures before and after the ceremony.

Their marriage was rich with cooperative efforts. They would take jaunts to scenic spots so she could paint and he could revel in nature. She illustrated *Three Little Folks, Idyls of Old New England*, and *The Hope of the World* with appealing line drawings. At home, she enjoyed gardening. Together they often sat on their porch, welcoming passersby, especially children, some of whom, now grown and living in the area, still fondly recollect them. He dubbed their place Bird Acre in honor of the "woodpeckers, nuthatches and chickadees" that he and Bessie began to feed as soon as they moved in.⁵⁶ He lovingly described the year-round convocation of winged pilgrims that visited the south side of their home, "a labyrinth of grape vines," where the couple provided seeds, crumbs, and meat at their "Bird Hotel" and were repaid by "the rat-a-tat of their bills upon the bottom of the boxes ... merry music, sounding for all the world like a smart shower upon a tin roof."⁵⁷ The two appreciated each tree around the house: "Next to the birds, the trees are my best nature friends at Bird Acre."⁵⁸

Most important, Bessie encouraged Hawkes, support he noted in his interviews and book dedications. After three decades together, he tenderly called her a "generous" woman: "my eyes, my right hand and my constant helper [who] has furnished much of the light of my life. I gladly give her credit for much of my achievements."⁵⁹ Five years later, in his dedication to *The Master of Millshaven*, he called her "My Dear Wife after whose courageous life, I have modeled the heroine of this book." (This noble character, Evelyn Henderson, daughter of a heartless tycoon, performs good deeds among the poor and eventually converts her father to philanthropy.)

Despite living in a small town in the state's west, and being only 30, Hawkes made himself known to a larger audience. In 1900 the prestigious Boston Authors' Club elected him to membership. His associates were older and famous. The president, Julia Ward Howe (1819-1910), had nominated him. Author of the famous "Battle Hymn of the Republic," she advocated pacifism, votes for women, and Mother's Day, all topics that resonated with Hawkes. Mother-in-law to Michael Anagnos, who directed Perkins from 1885-1890, Howe read to students once a week. She recalled Hawkes from those sessions and the two remained friends "For about twenty years."[60]

Hawkes continued to earn respect. After the death of his mother in 1899, he began to specialize in the nature works that made his considerable reputation. He initially published short sketches in publications like *Woman's Home Companion*, *Outing Magazine*, and *Overland Monthly*. These early vignettes predicted the longer works to follow. They looked directly at natural process, including pain or death, even when advertised as material "For Boys and Girls." A typical example: one chilly night a farmer wounds a grouse who suffers agony as it burrows in snow or hides in branches to evade a predatory fox and a hungry owl.[61]

Always a quick learner, Hawkes moved on from such modest compositions to create more than half a hundred full-length studies of, first, *Master Frisky* (1902), a dog. Then, like Noah's ark, his books transported frogs, beavers, birds, bears, moose, wolves, bison, "bronchos," reindeer, Shetland ponies, circus elephants, fox, cow ponies, circus horses, and turtles from their habitats to the eager imaginations of uncounted youngsters. His menagerie lived, as he proudly observed, in discrete sections of a wide universe: "Alaska, Arizona, Wyoming, upon the great central plains, in Canada, in Finland, and on the Malay Peninsula, not to mention north of the Arctic circle in Eskimo land."[62]

With buoyant enthusiasm, he carefully researched and interviewed to get his details correct. Uncle William Hawkes, who served as his secretary, had early regaled him with tales of the west as it was when he and father Enos explored it; later he would continue to interview travelers as well as check out armloads of books from Jones Library, Amherst, Goodwin Library, Hadley, and Forbes Library, Northampton. His brother Ernest William Hawkes became an authority on the native peoples of Alaska, perhaps inspiring such works as *Igloo Stories*. Along with unvarnished facts, he welcomed Mark Twain-like tall tales: a 1901 letter from the Ashfield town historian Frederick G. Howes informed him of a lightening flash that "knocked Payson sensible," of an Uncle Eli, who was told that if drafted for the Civil War he "should hire him a prostitute," of a day when "the thermometer was 121 degrees below zero," and of "one cow that was 17/16 Jersey."[63]

Almost immediately, editors of anthologies included Hawkes' writings. In 1904, two short dog essays, "Favorite House Dogs for Children" and "How To Manage Pet Dogs," appeared along with a photograph of Frisky, who resembles a Border Collie, in *The Children's Dog Book: True Stories of Home Dogs*.[64] Like his predecessor Dante, who imagined that famous pagan poets welcomed him among their company, Hawkes must have felt that the nationally distributed collections rewarded his decades of diligent composition. One needs to take a deep breath before naming his colleagues, current and deceased, in these compilations. Being grouped with luminaries from Shakespeare to Amy Lowell and offered to such large audiences certainly compensated him for those early, small Chautauqua groups in chilly churches or hot barns.

He forged personal as well as professional bonds with

some of the authors in the anthologies. His scrapbooks contain dozens of fan letters. He would send out a composition or book and regularly receive grateful thanks from notables. Hugh Lofting, for example, wrote from his Connecticut home to say that his daughter was enjoying Hawkes' novel *Wanted a Mother*. Perhaps with some embarrassment, Lofting admitted that his Doctor Doolittle books were never intended as "nature stories.... They are, as you will see, purely fantastic with just enough seriousness to make them plausible narrative [sic]."[65]

Hawkes' numerous appreciative fans picked up what they wanted from his works. Edgar Rice Burroughs, creator of Tarzan, read Hawkes.[66] By 1921, technical journals mentioned his contributions.[67] The anthropologist Grenville Goodwin, an expert on the Apaches who died in 1940 at the age of 33, ingested the mystique of the undomesticated. In his early teens, "he was devouring a steady diet of boys' books by the likes of Ernest Thompson Seton and Clarence Hawkes–romantic adventures with animals and native heroes set in the Far North and the Wild West, articulating a classic frontier paradigm of rugged self-reliance and outdoorsmanship."[68]

First known for his poetry, then for these wildlife accounts, Hawkes also won esteem for other genres. As early as 1905, an article with a picture of him in his Hadley study applauded him: "Clarence Hawkes is essentially a writer of nature studies, though he is equally brilliant in fiction. Though blind, Mr. Hawkes sees with the eyes of an imagination that suggests soul memories of the years ago."[69] A 1915 ad for books in *The New York Times* paired his autobiographical *Hitting the Dark Trail* with Robert Frost's *North of Boston* and *A Boy's Will*.[70]

Popularity led to academic recognition from four colleges. He received the only honorary degree, a Master of Arts, at William Smith's sixth commencement on Monday, 11 June 1917. Because of the war, the ceremony for the 21 women graduates in Geneva, New York, was simple. Yet Hawkes moved the audience by telling a fable about a prince who "met with an accident, which made him unfit for the conquest of arms, but then he became a jester and by his wit, goodness and kindness he accomplished more than he could have accomplished by the sword." Respecting people who overcame "seemingly insurmountable obstacles" was exactly the right message because the school had already educated Elizabeth Blackwell, the first American woman physician.[71]

Syracuse University likewise conferred a Master of Letters on him on 11 July 1917.[72] Two years later, *The New York Times* reported that Amherst College had awarded an honorary Master of Arts on the previous day.[73] The gracious presentation said, "To a neighbor who has triumphed over heavy hindrances, who has read with his hands when his eyes failed him, who has taught little children the ways of Nature and older children the lessons of beauty and courage–to a neighbor who has won merited fame abroad we extend the hand of fellowship at home."[74]

Hawkes' last degree reflected his growing reputation. American International College in Springfield, Massachusetts, awarded him an honorary Doctor of Letters on 7 June 1938. The extended citation by the Reverend

Robert Merrill Bartlett of Longmeadow's First Church of Christ repeated a biography that had already inspired millions of people. Bartlett mentioned the accidents, the many books, the connections with civic life, and then reminisced about a conversation he once had concerning the author's obvious trips to Arizona, Nevada, and Alaska. Laughing, Hawkes amended the compliment: "I have never been west of Hobart College in New York state."[75]

Recognition continued throughout Hawkes' life. Here are only two admiring examples: *The Baltimore Sun* compared his career with that of Henry Fawcett, "the blind statesman, economist and Postmaster-General" of England. This pairing supported the author's claim that people with physical defects benefit from a "coefficient of resistance," a tension that urges them to accomplish more than people without impediments.[76] And on the fiftieth anniversary of his blindness, *Nature Magazine* quoted a eulogy by his late friend William Temple Hornaday (1854-1937), first Director of the Bronx Zoo, author and crusader to save American bison, Alaskan fur seals, and birds whose plumes adorned women's hats. Hawkes was

> the born Nature-lover, the woodsman, the chronicler, and the painter of mental pictures, who for a brief few years looked into the pulsing heart of Nature, focused his mental camera upon her during a few brilliant days, and the suddenly, with a stroke of lightening, the world became black.... With marvelous fidelity, he paints what he has seen and yet remembers.[77]

Although Hawkes rightly won respect as a poet and environmentalist, the epithet "author" unjustly narrows our appreciation of his many achievements. His involvement with sports, community affairs, and politics shows how many projects his disciplined mind accomplished.

He loved baseball throughout his life. After the boyhood amputation, he still "learned to make his wits piece out his legs, and so make up for his shortcomings. If he could pitch so cleverly that the other boys could not hit the ball, he did not have to field it."[78] In 1916 he took pride in "following the game nearly 30 years, ever since I was a small boy." While at Perkins, he first became aware of big league teams ("The Bean Eaters"); later he regularly attended games at Massachusetts Agricultural College (later University of Massachusetts), Amherst College, and Hopkins Academy. "By close attention, I learn from the sounds where the ball is and what takes place on the diamond."[79] His motor trip to New York City the next year allowed him to attend a game at the Polo Grounds when the Giants beat St. Louis 2-1.[80] Closer to home, he urged a competition between Northampton and Greenfield, reasoning that rivalry in sports promotes camaraderie.[81]

Football, too, intrigued him. Although he had never seen an actual game, other people described the action to him and he used "his own remarkable ability to change sound waves into light waves."[82] Impressed by his judgments, people asked him to pick players for championship teams and happily listened as he provided statistics to explain his choices. The football game in *Uncle Billy* provides an occasion to reform otherwise delinquent youngsters.

In addition, Hawkes enjoyed other pastimes. A laudatory article during 1936 pictures him placidly angling from the stern of a rowboat, perhaps in the North Hadley Pond, one of his favorite haunts.[83] At home, he avidly played cards and board games with specially constructed equipment. A rotogravure photograph from around 1936 shows him celebrating his 67th birthday by playing cribbage with neighbor Harry E. Bicknell. Both men wear gentlemanly coats and ties but each obviously aims to come out ahead on the scoring block between them.[84] His checkerboard, with specially raised markers, currently reposes in the Hadley Historical Society building on Middle Street. These activities remind later generations of the importance once placed on them: the 1900 Paris Olympics offered competitions in fishing and checkers.

Hawkes' pleasure in festivity benefited the 1909 celebration of Hadley's 250th birthday. Never having seen a parade float, he designed twenty of the forty ornate display wagons to remind watchers of the town's past. They carried tableaux of legendary figures like General William Goffe, "The Angel of Hadley," who supposedly saved the town from an Indian attack in 1675, and Molly Webster, accused of witchcraft. One displayed an enormous dictionary, reminiscent of the book given a prominent place in his West Street study. To decorate these flatbeds, carriages, and autos, he superintended the manufacture of "thousands of beautiful crepe paper chrysanthemums, roses and poppies. Some 30,000 people watched the extraordinary procession of nearly 700 participants and 200 horses, "reaching two miles, and taking an hour to pass a given point."[85] *The New York Times* judged it better than big city affairs and marveled that the impressive spectacle "was entirely the work of a blind man."⁸⁶

Hawkes' politics recalled his early familiarity with poverty. He favored leaders like Herbert Hoover over Franklin Delano Roosevelt, reasoning that those born in humble circumstances would be more sympathetic to common people than those who never were poor. Praising the late Warren G. Harding, Hawkes recalled how the president came from a large family, struggled like Grant, Garfield, and Lincoln for an education at "a small American college of the Middle West," boldly plunged into business, and urged a shorter workday. These traits, plus

1909 ANGEL FLOAT

his "friendliness," may help subsequent citizens "learn how to live."⁸⁷ Coolidge, an early admirer of Hawkes, received similar approval in a much reprinted essay called "Granite and Gold." Recalling his own boyhood, walking "three

HAWKES' CHECKERBOARD

miles to the nearest village" "through a dark forest" (like Dante) past a quarry, Hawkes fancied there was gold amid the rock. He applied the image to Coolidge, inflexible in principle, minimalist in speech, a healthy model for people because "Never in the memory of man have the ideals of humanity been so shifting and evanescent."[88] Coolidge had not abandoned his Pilgrim and Puritan rectitude, so Hawkes positively called him a "Super-Yankee."[89] The same admiration for stability prompted Hawkes to stand by Hoover during the 1932 election. An orphan, son of a blacksmith, Hoover knew poverty ("Born in a home so humble that Christmas was made glad only by popcorn balls and hickory nuts"). "He is preeminently a self-made man and he did a mighty good job, one that money-made candidates can never know."⁹⁰

 Cynical modern students of politics may question the validity of Hawkes' tributes. Like all leaders, the ones he commended had flaws, but he chose to ignore them. Just as he disliked jazz and preferred long-established music, so he looked backward for heroes. While some of his younger contemporaries rejected what they sneeringly called outworn Puritan attitudes, many like Hawkes valued tradition. He was not alone. A host of conservative authors paid little attention to the social sarcasm of H. L. Menken, the depressing portraits of country life by Edgar Lee Masters and Hamlin Garland, or the allure of F. Scott Fitzgerald's pleasure obsessed flappers and playboys. Defenses of times gone by ranged from nasty—the eugenics movement that worried about mixing pure Nordic peoples with lesser races—to reasonable to sentimental.

 Hawkes moderated his preference for proletarian leaders when, in "A True Fish Story," he recounted a conversation his friend had some twenty years before. On a lake at Cape Cod, Fairman (the justly named narrator) encountered two fellow fishermen, one plump and the other "slight and dapper." Later on a train the three meet again and the small man asks Fairman's opinion of Grover Cleveland. With words like those of Hawkes, he begins with admiration ("Well, … as a self-made man, one who has come up from the bottom of the ladder without much pull and with few strong boosters, I admire him"); soon, though, he forthrightly critiques the President's performance—only to learn when the duo leave that they were Cleveland himself and his advisor Joe Jefferson.[91]

BESSIE AND NIGHTWATCHMAN

Hawkes' political opinions synchronized with his concerns that time and human interference could diminish nature. Birds, beavers, fish, and forests will survive only when people restrain their innate rapacity. As a young boy, he copied his father and hunted without reflection: "If I had possessed some of Thompson Seton's or G. D. Roberts's books, I might have seen another side of the sport."[92] Later he feared that over-killing wildlife or altering land would depopulate the countryside, a realistic anxiety born out by the extinction of the passenger pigeon and the near disappearance of bison and beavers. Once again, Hawkes aligned himself with far-sighted naturalists like Henry David Thoreau, who grieved that humans barred fish from the Concord River: "Perchance, after a few thousands of years, if the fishes will be patient, and pass their summers elsewhere, meanwhile, nature will have levelled the Billerica dam, and the Lowell factories, and the Grass-ground River run clear again."[93]

Most important, Hawkes' writings harmonized two conflicting views of the external world that had energized strong debate, especially among Americans. William Wordsworth encapsulated the belief of the idealistic school by urging, "Let Nature Be Your Teacher,"

> One impulse from a vernal wood
>
> May teach you more of man,
>
> Of moral evil and of good
>
> Than all the sages can.[94]

By contrast, William Bradford, just arrived in the new world after a horrendous sea voyage with his Pilgrim émigrés, abhorred the wintry Massachusetts landscape of 1620 as an impediment to spiritual growth: "[W]hat could they see but a hideous and desolate wilderness, full of wild beasts and wild men…. [T]hey could have little solace or content in respect of any outward objects."[95] In Hawkes' time, the simple contrast between nature as divine and nature as the enemy of civilization had evolved. Early nineteenth century patriots praised our country's sublime scenery because it proved we were God's favored children; in the latter part of the century, acquisitive capitalism forced citizens to admit that everywhere they were cutting down these same ancient forests, planting non-native crops on the ocean-like prairie, exterminating unique species, dynamiting the awesome mountains, and damming up pristine rivers.[96] Hawkes' 60-some nature books honor both visions, giving dignity to the not-human and, simultaneously, regretting the danger it poses when civilization progresses.

    The Connecticut River flood of 5 November 1927, which forced him from his home on fifteen minutes' notice, prompted him to think of how people could benignly assist in the grand scheme of natural process. The previous flood in March 1913 had left ice mounds on Hockanum Road "higher than a horse's back"; this latter flood caused "great loss in tobacco ... onions ... carrots," mainly because little had been done to channel the river's violence.[97] Hawkes suggested an ambitious plan to preserve the environment. Women over the age of fourteen should plant at least one tree per year, place and maintain one bird house, vote for conservation laws, and protest any interference with best management practices. His appeal to women invoked a higher value than fashion, namely civilization itself: "[B]anish the adorning of hats with birds' feathers to the age of barbarism. The painted savage adorns his head with birds' feathers, but the refined white woman should know better."[98] It seems inappropriate now to criticize this expression as politically incorrect; in its day, it offered a reasonable solution to a real problem by means of a powerful and almost universally accepted choice. Perhaps he privately honored grandma Gurney ("it was of her that I learned to feed the birds"; "she was a defender of wild life").[99] Widely reported in newspapers after March 1928, the suggestions for a female conservation corps took root in some schools, especially out west and at Holyoke High School.[100]

These have been the major events of Hawkes' exemplary life. However, happenings that seem minor may help to understand his personality. Once he mourned the shooting of his little fox terrier on the Hadley Common. An often reprinted article chronicled Foxy's value: listening with Hawkes when Amos' dog barked on radio's *Amos 'n' Andy*; begging scraps from the table; snuggling on his lap; and keeping Bessie company when she peeled potatoes.[101] Another time he speculated that animals might have a future life.[102] After scolding vivisectionists, boys who stone dogs, and careless drivers, he listed the benefits of a canine companion that deserved reward ("Fidelity, loyalty, trustfulness, honesty, devotion, selflessness") and hoped "to Be Worthy of His Dog."[103] Obviously, animals demonstrated the love that he sought to promote. When Sarah Berhardt's Airedale ran away while she performed in Northampton, it arrived on the Hawkes' doorstep. Although he returned it physically, it survived as the loyal helper of his 1924 *A Gentleman from France: An Airedale Hero*.

Hawkes died from heart failure a month past his 84th year at 8:30 a.m. on 19 January 1954 at Northampton's Cooley Dickinson Hospital. Bessie survived until a heart attack took her in her 88th year on 28 January 1958.

Although their physical hearts failed, their spiritual hearts were always strong, inspiring others to strive and seek and not to yield. Today they rest together with the first settlers of their town in the peaceful Old Hadley Cemetery, between the newly cleaned Connecticut River and fields still plowed in narrow medieval strips, a survival of the past that may be unique on earth today. They always did enjoy continuity, human and natural.

HAWKES AND HIS TYPEWRITER

The Writings

Poetry

As the famous naturalist Dallas Lore Sharp realized, "None of Mr. Hawkes's books need a commentary and affidavits. His purpose is plain, his story simple, his language clear."[104] His poems were public communication, not modernist self-indulgence. Both in technique and subject matter, they looked backward. He candidly announced his standards:

> I am something of a reactionary all along the line. I do not care especially for free verse. I have always considered it a lazy man's way of writing poetry. When a poem is not metrical, has no color or sense, why should we label it poetry?... Perhaps I am a back number, but I much prefer Tennyson and the Brownings to Swinburne and Ella Wheeler Wilcox.[105]

Unlike the hermetic utterances that learned authors like Ezra Pound and T. S. Eliot and Wallace Stevens delivered to a new generation of readers in the early twentieth century, Hawkes' words expressed universal impressions, artfully expressed, yet determinedly accessible, all in the manner of popular nineteenth-century verse. He proudly strode up the mountain of poetry in the company of acknowledged models such as William Cullen Bryant, Henry Wadsworth Longfellow, and John Greenleaf Whittier. Today, even large anthologies of American verse barely mention these teachers, but in their day they amused, instructed, inspired, and impressed uncounted readers. To read them guaranteed that one understood them. Eager for recognition, Hawkes did not invent a flamboyant persona or experimental language. Rather he obeyed Alexander Pope's advice for poets in 1711: write "What oft was Thought, but ne'er so well Exprest."[106] The result of his perfective imitation still can engage open-minded audiences who do not demand continual novelty.

Most of his poetry is (in the phrase of critic Kenneth Burke), "equipment for living,"[107] because it describes encounters with the physical world and then explains the consequences of such contacts. In the words of Old Ben from *Stories of the Good Green Wood*, who quotes Shakespeare's Duke in *As You Like It*, Hawkes' verse finds "books in runnin' brooks, sermons in stones, an' good in everything." (48).[108] His speakers, their settings, the subject matter, and the poetic techniques sum up a century of vibrant populist verse.

Repeatedly, especially in the five early collections, Hawkes' speakers promulgate the Rousseau-Wordsworth-Whitman respect for common people. Many narrators talk in unlettered dialect but are the authority figures. The first page of *Idyls of Old New England* uses "nater" for nature, "an'" for and, "ain't," "'ud" for would, "wa'n't" for weren't, "o" for of, "yer" for your. These colloquialisms further the notion that plain speaking better conveys honest emotion. The narrator of "Ma's Posy Beds" admits,

> Somehow I like the good ole kind
>
> O' posies full as well,
>
> (Instead o' those with Latin names
>
> That nobody can spell,)
>
> Like mirigold an' asterziz
>
> An' calliopsis fair...[109]

Even when the verse concerns celebrities like Hadley's Civil War general Joseph B. Hooker, the language remains simple. One of Hawkes' most famous poems imagines a black Southerner named Ole Moses as he recalls "How Massa Linkum Came" to Richmond after the Civil War. A preacher prays with the President in front of the crowd,

> An' den dey broke into a shout
>
> Dat mought hab woke John Brown,
>
> An' cheered until I t'ink de noise
>
> Would bring de heabens down.[110]

Far from demeaning this engaged speaker, the unlearned lines communicate his reverence for our iconic hero and further the optimistic wish of Henry Grady and

Hawkes' graduation speech that the war was worth the losses. Lincoln's son enjoyed the poem.

The speakers in *Three Little Folks,* his second volume, are children. Perhaps like his own brothers and sister, they share their "sweet and fanciful stories" with the poet, "the old man who lives down in the village." He writes their experiences "in a book," a plot device Hawkes would later use in his novels.[111] The rural accounts of "A Merry-Go-Round" made from cart wheels, "Sliding Down Stairs" on Papa's chair (which shatters), quoting "What the Brook Told Me," and "Playing House" present a uniformly positive panorama of rural life. As one reviewer perceived, the language fits the young speakers and does not merely express adult concerns: "he does not utter as the voice of childhood those mysterious thoughts which are as a rule the readings of the overtones and undertones of older years into the plain song of simple life."[112]

When narrators address readers in received standard English, they emanate the same earnestness. Two stanzas from *The Hope of the World* call attention to earthly decay and suggest a pious remedy:

> Is this sad wall, of dark and crumbling stones,
>
> That time hath eaten like a dead man's bones,
>
> The bulwark of old Rome who ruled the world
>
> And builded empire from a hundred thrones?...
>
> O let Christ's spirit through the world awake,
>
> Let hands reach down and hands degenerate take,

> Let shoulders strong bear burdens for the weak,
> Let love abide e'en for the whole world's sake.[113]

A literary analyst could call attention to the unobtrusive technical devices: the AABA rhyme scheme popularized by Edward FitzGerald's epigrammatic *Rubáiyát of Omar Khayyám* (1859); the first stanza's erotesis/rhetorical question; the adjective "sad" transferred from object to viewer; the prosopopoeia/personification of time; the recordatio/summing up past events in stanza one, lines 3-4; the anaphora/repetition of "let"; the aureate diction of "O" and "e'en"; the synecdoche of "hands" and "shoulders," which mention only part of a whole; the parallel structure of stanza two; the invocatio/prayer form of stanza two; the whole sequence as both an apocarteresis/leaving one hope while turning to another and a philophronesis/attempt to stop God's wrath by humble repentance. But such misplaced learning, although common in high schools and colleges of Hawkes' day, contradicts the demotic spirit of his diction.

As with speaker, so with setting and subject. Hawkes places us in a tactile world in which common objects like hayfields or starry nights teach us our place in the physical, social, and spiritual realms. This conviction that all nature is a system of signs which, rightly understood, can improve us unites mystics from every culture.

> I love ter think we humans grow toward God
> Jest like the lily rises from the sod,--
> That friendship with the fields an' mother earth

> Give ter the human soul a wondrous worth.[114]

Sometimes the object appears only as a springboard for moral decoding:

> Don't let no hole git in yer moral fence,
>
> But keep it jest above Temptation's nose,
>
> For if one little peccadillo goes
>
> Ter tother side a score will follow hence.[115]

Other times the speaker emphasizes a concrete phenomenon in order to discover its lasting importance. Several poems discuss the sinking of the battleship *Maine* in Havana harbor. No matter what the cause, Spanish treachery or accident, chaos affected everyone.

> My God! That flash, that flame, that deaf'ning roar
>
> That fills the vaulted heavens o'er and o'er!
>
> That sudden shock that turn the peaceful sea
>
> Into an Aetna's dire ferocity.[116]

Despite the lurid possibilities of the incident, which "yellow" newspapers exploited, Hawkes quickly goes past the emotions it inflamed to extract an ethical meaning.

> Remember the Maine, but not with bitter hate
>
> Against the land that may have done her wrong,
>
> For God will judge although He tarry long,
>
> His arm is sure, His justice is not late.[117]

Peace will come after the literal explosion of the ship and the metaphoric outburst of war to insure "man's ire. / The Rage of war, the venom battle brings, / Melt like miasmic mists, when morning flings / Her beams afar."[118] The lesson of his meditations on the blast reflects the missionary zeal of his era: "Let liberty and justice be your aim, / In darksome lands let truth's bright beacon shine."[119]

In short, the palpable universe with its frogs and waterfalls and passing seasons pleases on two levels, the instinctive, sensory one and the educated moral plane. Everything terrestrial exists for our delight and our edification, the perfect work of human art envisioned by the Roman poet Horace, but created by a compassionate deity. It upholds a democratic philosophy by demeaning cities and college snobs and wealth while elevating sunshine and flowers and birds, humble chores like plowing a field or making maple syrup as well as stern ancestors and heartfelt religion: "[I]t wuz a rugged gospel / That inspired our noblest deeds."[120] His Christmas and Easter poems reiterate the possibility of universal fellowship.

We can only guess how readers treasured Hawkes' many poems. One tribute may stand for many others. Clara Clough Lenroot recalled the hard life her mother lived on a farm near Osceola, Wisconsin, near the St. Croix River in the 1860s. When she died, Lenroot, who had recited Bryant and Longfellow on her two-mile walk to school, found the most appropriate tribute for her departed parent in the words that praised mother Edlah Hawkes: "I am reminded of the lines by Clarence Hawkes: TIRED HANDS," which she quoted in full.[121]

James A. Freeman

Nature Works

Shortly after Hawkes' mother died in 1899, he shifted his attention from poetry to narratives about animals. Like other Victorians such as George Meredith and Thomas Hardy, who also redirected their energies, he threw himself into the new interest and became something of a cultural revolutionary. He boldly challenged two commonplace ideas. First, he described nature as a self-contained entity that was not primarily an escape destination for males who did not wish to grow up or as a realm of mystic transport. Next, he reconciled two seemingly incompatible ideas about nature, that it was almost divine and that it was demonic. (See Appendix: Nature Writing Before Hawkes.)

Almost any observer could have predicted that Hawkes would write truthfully about the outdoors and thus often contradict fashion. "I had been born and bred a child of Nature," he recalled. "There was no sound or scent of hers, no song, or sound of woe, that I did not know."[122] From his first rambles in the woods with his father, from his earliest guidance by his mother and by grandma Gurney, he felt that all living things possess an identity: they sing a unique song, indent a distinctive footprint, sprout differently shaped leaves, or live in artfully adapted dwellings. His memory of phenomena provided the basis for a lifetime of clear observation: "[I]n some mysterious manner," he said, "the sensitive plates exposed in my youth secured perfect pictures of each passing season and of a multitude of the varying aspects of nature."[123]

Hawkes' encyclopedic reading augmented his capacious memory, providing him with narratives from two millennia in which four-footed or two-winged creatures reveal truths about human lives. From the Greek slave Aesop, whose turtle always outraces the hare, through medieval cycles like that of Reynard the fox, who constantly tricks Isengrim the arrogant wolf, to Uncle Remus' Brer Bear, Brer Fox and Brer Rabbit, the menagerie had amused and instructed generations. Although the stories' purposes varied–ethical training in Aesop, peasant criticism of the predatory 12th century clergy, preservation of black folk tales after our Civil War–they all justified a second look at the non-human world.

Hawkes' early stories understandably glanced backward to models like Aesop. Old Ben Wilson, who instructs the young narrator in *Tales of the Good Green Wood*, retold a fable "how 'twas the rabbit lost his tail"(9) to the waddling-but-persistent turtle. Likewise, "A Night with Ruff Grouse" quickly established the danger faced by the wounded old partridge by invoking the ancient title for a relentless adversary: "Ruff's feathers stood up with fright and his eyes grew big with terror: it was Sir Reynard, and he was after him"(49). The popular dialect fables of Uncle Remus probably encouraged Hawkes' evocations of local color characters like Old Ben.

Hawkes repeatedly demonstrated how raccoons, plovers, weasels, and circus ponies have unique traits that should be appreciated. They live interesting but non-human lives in a beautiful yet dangerous world. His realistic biographies admit they fight for survival, clothed in fur or feathers, not spats, but also love their offspring, enjoy good weather, exhibit curiosity, can be loyal to deserving partners, and may remember their past actions. A few times in his books, a beast resembles a person (polar bear cubs "wrestle like two boys"[124]). Usually, though, his first concern is to accurately depict the life cycle of his subject without imposing on it the pattern of human development or emotions. Gene Stratton Porter accurately observed, "How Mr. Hawkes does it, I do not know, but he does describe nature sympathetically and accurately. His animals are not humanized. He has the wisdom to recognize that the processes of nature are distinctly cold-blooded."[125]

He allows us to see how a sled dog lies on his back after a long day's haul and chews off the ice that has congealed on his paws; he marvels that "a young robin requires 14 feet of earthworms per day"[126]; he informs us that the half-digested moss taken from a slain reindeer's belly, mixed with berries, provides hunters in Lapland with a tasty meal[127]; he takes us on a tour of a beaver's lakeside dwelling, noting the materials, dimensions, and reasons for the design.[128]

A few of these data may make a reader smile, as one does when reading about Kenneth Graham's anthropomorphic Toad from *The Wind in the Willows*; most educate; some may prompt a change in moral position about interfering with nature. All, though, transport us to the green woods or snow-swept tundra so often forgotten by unobservant, urban humans. They substitute observed truth for that of irrelevant cuteness or lurid savagery. Further, they invite readers to appreciate a plenum, a universe filled with fascinating activity. The speaker of *Big Brother* reminds readers that in the wild, animals have passed, are passing, and will continue to pass whether or not anyone observes them:

> There were many tracks in the snow on this winter afternoon that interested Red Fox. There was the two by two of the red squirrel, and also the two by two of the weasel. These two tracks were identical, save for the fact that if one looked sharply he might have discerned where the hair on the weasel's belly had brushed the snow between jumps. The T shaped track of the rabbit was unmistakable, with the shank always pointing in the direction the rabbit is traveling. The scraggly tracks of the partridge were for all the world like the footprints that the barn fowls make in the farmyard. There was no mistaking the lace-like track of the wood mouse, and the clumsy trail of the skunk was like no other rail; often his pudgy belly fairly dragging in the snow.(12)

The narrators in his books often examine one animal but then shift focus to include others, always aware that innumerable living creatures inhabit our globe. A year-old walrus, minutely described in *Igloo Stories*, lives in a frozen environment that also houses "The sea-cow and the sea leopard, …gulls, … little auks, …White Arcticgeese [sic], … Blue fox, …polar bear, …killer whale" (45-46).

Hawkes' attention to nature seems so visceral that no one would accuse it of being a mere retreat from adult complexity. When Hawkes spotlighted people, he seldom made them conform to the older escapist pattern of many American narratives in which one or two males fled to a wilderness, mainly so that they could indulge their self-centered fantasies. Washington Irving's henpecked Rip

Van Winkle set the pattern for James Fenimore Cooper's Leatherstocking and Chingachgook, Herman Melville's two young sailors who jump ship onto a South Pacific island in *Typee*, and Mark Twain's Jim and Huckleberry Finn who (in the latter's words), "got to light out for the Territory ahead of the rest, because Aunt Sally she's going to adopt me and sivilize me."[129]

    Hawkes bravely diverged from this trendy flight pattern. He eagerly entwined himself with an authentic natural world to augment his relation with all life, not separate himself from any part of it. When he does depict a mentor-pupil encounter, he emphasizes the vital knowledge that connects both the older generation and the younger in contact to an eternal system. In *Big Brother*, for example, the aptly named hired man Uncle Solomon teaches the farm owner and his son to recognize the tracks of a marauding bear, to attract him by leaving a dead sheep in the open, and finally to shoot him from a skillfully constructed blind (58-68). Personally, Hawkes used his knowledge of the physical universe to bond with other people. He committed to memory around 80 separate birdcalls, using records sent by Cornell University and the New York Foundation for the Blind,[130] and called his West Street home Bird Acre. Intrigued by this closeness to birds, Thornton Burgess wrote him a technical letter asking the exact kind of grebe— horned, Holboell's or Western—he had noticed in Hadley. Burgess promised to read Hawkes' reply over radio stations WBZ (Springfield) and WBZA (Boston).[131]

    His books carry on the habit of unobtrusive observation. The most vibrant interchanges take place between wild creatures acting instinctively: Old Ringtail the raccoon, although raised by a boy, kills more hens than necessary, steals vegetables from a garden, and fights to the death against a big coon dog.[132] If there is a human eyewitness, this onlooker would presumably mature after learning about animal instinct and take the new insights back to a human community.

Hawkes' stories prudently accepted nature as both inspiring to the human soul and potentially lethal. They rejected binary extremes that forced a reader to decide whether the world will comfort or crush people. On the one hand, they confirmed a Romantic inclusiveness by sponsoring the benefits of contact with creatures. He fixed his inner eye upon the features anyone can observe, all the while implying that such clear vision will promote in the audience feelings of at-homeness in the world. Coming close to genuine animals, people inevitably benefited. As the narrator in *Dapples of the Circus* says,

> In these days of machines, when so much is artificial, what better playmate could children have than one of these little horses? Learning to care for him and drive him under all conditions, develops character as almost nothing else will. (18-19)

Readers who had once experienced unmediated contact with the wilderness would find a parade of Benjamin Bunny, Peter Rabbit and Jimmy Skunk patronizing.

On the other hand, Hawkes admitted even the most deserving humans could expect little direct help from nature. In *King of the Flying Sledge*, when Olga, daughter of old Oscar and husband of Hans, lies dying from childbirth in frozen Lapland, her father passively murmurs, "We must have faith; God is good." Hans, however, sets out by sledge to fetch a doctor. On the trail he marvels at the Northern Lights. True, the "orange, crimson, blue, yellow" sky revealed (in the words of Psalm 19) God's glory and handiwork, but they did little for his needy wife and child:

> Just as though the Creator was setting off some mighty fireworks, or pyrotechnic display, to awe the antlike creatures who crawled to and fro on Mother Earth and called themselves lords of creation. This was the thought of Hans as he gazed reverently over his shoulder. How insignificant and puny we were, after all, when face to face with the workings of nature! How little the elements, the wind, fire, and water, considered man when they went mad and did the work of nature! (19, 27-28)

Despite acknowledging the basic harshness of nature, Hawkes did not magnify it. One comparison illustrates his non-sensational approach. Jack London's "Bâtard," the story of the ornery sled dog Diable, revels in sadism.[133] Not only does Diable fight with dogs and men, but his cruel master, Black Leclère, starves and beats him. To many animal rights activists, such cruelty reflected badly on human who practiced it. Readers were given no reason to excuse similar beatings administered to Ouida's dog of Flanders by a sadist who leaves the loyal animal for dead because it fails to work for him. London implies a kind of justice in his frigid universe, but one that elevates nobody. At the end, Leclère, accused of murder, is about to be lynched. As he stands on a box with a noose around his neck, news arrives that the real killer has been found. The vigilantes leave, cruelly assuring him they will come back eventually. Before they return, Diable knocks Leclère from his support and, when the posse returns, must be shot to loosen his grip on the dead man's leg.

By contrast, Hawkes' sled dog Silversheene suffers similar abuse from the Canuck François Dupret, "a gambler and a notorious Alaska bad man."[134] But the noble dog gets his revenge by pulling their sledge into a crevasse, thus trapping the unworthy master. With cool detachment, the narrator tells how later "a pack [of wolves] had descended upon the scene of the disaster to François and his team and picked their bones clean" (124).

Hawkes' vision of nature as simultaneously healing and destructive might have offended contemporaries who mentioned most facts of reproduction and survival indirectly. However, he calmly recorded births and deaths. For instance, *Wood and Water Friends* has a familiar introductory note:

> This is God's world that we live in. With His hands He fashioned it; the woods and the waters. He made it for an inspiration and delight. So let us be mindful of its beauty and its wonder and hold them in our hearts as a sacred precious thing that shall make us rich as Kings and fill our souls with peace and understanding. (xii)

But the soothing piety coexists with Darwinian objectivity: in the forest, "no creature ever dies of old age"(29); despite other writers' fantasies about animal parliaments, "Life is a battle"(203); all obey "that instinct of self-preservation"(259) that directs them to kill. In this one book Hawkes mentions the deaths, usually violent, of a sparrow, woodchuck, ferret, baby woodpeckers, baby robins, snake, hawk, weasel, hunting dog, frog, baby fox, otters, fish, muskrat, baby jay, mink, and bittern. The descriptions vary from matter-of-fact to this lyrical account of a hunter dispatching a mink caught in his trap:

> Then he raised his club, which descended swiftly, and the song of the Little brook was stilled in the Terror's ears, and he swooned away into breathless darkness, and was nothing but a sleek pelt." (287)

When Hawkes shifted his attention from animals to their environment, he often forced readers to confront landscapes that had been altered by man. Pep, the brave bull terrier, searches for his wounded master in the Argonne forest during the World War. Later, Lassie-like, Pep will bring aid to him, but the faithful dog must first travel through a wasteland:

> The region had been the scene of heavy fighting for two days, and the signs of war's horrible devastation were on every hand. Shrapnel had stripped the trees of much of their foliage. Many of them were down while others were torn and broken, with limbs hanging or strewed on the ground. The whole face of nature was scarred and furrowed, seamed and made hideous by the passing hurricane of battle.[135]

Hawkes' sorrow for "the fair face of France" (91) differentiates this passage from one published ten years later by Ernest Hemingway. In "A Way You'll Never Be" (1932), Hemingway's shell shocked American volunteer Nick Adams rides his bicycle past another battlefield in Italy and for five paragraphs lists the man-made objects that register on his consciousness:

> [The dead] lay alone or in clumps in the high grass of the field and along the road, their pockets out, and over them were flies and around each body or group of bodies were the scattered papers.... [T]here was much material: a field kitchen, ... many of the calf-skin-covered haversacks, stick bombs, helmets, rifles,

> ... intrenching tools, ammunition boxes, star shell pistols, ...a squat, tripodded machine gun ..., the crew in odd positions, and around them, in the grass, more of the typical papers.[136]

Both authors lament the disorder caused by combat, but Hawkes concentrates on the unseemly mutilation of trees, and, by extension, of all nature, while Hemingway communicates horror at the misuse of human technology. The wounded soldiers whom Pep sees during his rescue mission certainly move us, but Hawkes conveys emotion by the powerful image of a disfigured wood, not their equipment.

Hawkes forced his audiences to accept their involvement in the natural world. One of his *Notes of a Naturalist* sounds an ominous warning based upon his lifelong study:

> It is the fact that the wild kindred war so continually on one another. The larger on the smaller, the stronger on the weaker. Life here in the ancient woods is a continual struggle. I suppose that his law helps the species, for only the strongest and wisest survive, but I fear that in the immediate future a great plague is coming on the world because of the folly of man. (77-78)

His credibility made such a jeremiad powerful. He had escaped the "Nature Faker fuss" of the century's first decade, a battle for authenticity he outlined in *The Light That Did Not Fail* and *My Country*, and waged with each composition he published.

By any standard, Hawkes' many nature books qualify as genuine environmental documents. To earn the title, one checklist requires that:

> The nonhuman environment is present not merely as a framing device but as a presence that begins to suggest that human history is implicated in natural history ...

> The human interest is not understood to be the only legitimate interest ...
>
> Human accountability to the environment is part of the text's ethical orientation ...
>
> Some sense of the environment as a process rather than as a constant or given is at least implicit in the text ...[137]

A second checklist enrolls Hawkes among the ranks of "ecocentrists" because he agrees with the following creeds:

> —natural systems are the basis of all organic existence, and therefore possess intrinsic value
>
> —humankind is an element within rather the reason to be of natural systems, and is hence dependent on intrinsic value
>
> —ethical human actions (actions which promote the good life for humankind) necessarily promote all life on earth (preserves such intrinsic values as diversity, stability, and beauty).[138]

Hawkes' appreciative readers can give copious examples to demonstrate how he supports each of the above requirements. Here it is enough to say that his writings bravely contradicted popular hypotheses about nature, notably that it primarily beckoned men unable to cope with society and that it either inspired or destroyed.

Novels

Hawkes wrote four novels that focused on people rather than animals. Still, each reinforced the message in his poetry and nature books: we are inextricably connected to the world whether or not we admit the bond. *Wanted a Mother* (1922), *Doctor Thinkright* (1934), *The Master of Millshaven* (1936), and *Uncle Billy, the Curious Cobbler* (1939) all asserted that self-destructive narcissists can be converted to joyful members of a larger society by means of a dramatic encounter. This experience of being morally "inverted" or "turned bottom side up" (as the spunky heroine of *Wanted a Mother* puts it [250]) recalls two of the oldest Christian success tales. Saul the Jesus-hater transformed into Paul the missionary after his unhorsing on the Damascus road. Later at Milan during the summer of August 386, the spiritually tormented Augustine sat weeping under a fig tree in the garden of his cottage. Miraculously, he heard a child's voice chanting, "Pick up and Read!" The biblical text he chose changed his identity from libertine to truth seeker.[139] Hawkes' novels added to these standard conversion scenes Victorian narrative tactics to encourage readers that it is possible to transform oneself into a better person if one heeds the benign signs in nature, in human nature and in literature.

Wanted a Mother presents a threesome of characters familiar from well-read nineteenth-century fiction. The eight-year-old orphan Eleanor Abbott leaves the Ashton poor-farm to live with her aunt Lucretia and uncle Nathan, the late mother's sister and brother. Eleanor sums up a century of worthy but abandoned children. Charles Dickens' *Oliver Twist* (1839), *David Copperfield* (1850), and Pip of *Great Expectations* (1861) joined Charlotte Bronte's *Jane Eyre* (1847), Wilkie Collins' Madonna Grice of *Hide and Seek* (1854), George Eliot's Esther Lyon of *Felix Holt, the Radical* (1866), and Ouida's Musa of *In Maremma* (1882) are the best-known waifs. Eleanor has both the beauty—blond curls and big brown eyes—and the feisty truthfulness that moved audiences. All she craves is someone to love her.

Like her literary predecessors, however, she finds the people nearest to her cannot supply effective affection. The cold Lucretia (perhaps a descendant of the feared Lucrezia Borgia) humiliates her niece because Eleanor's mother had married the man she loved. Although kindly, Nathan can do little to mollify Lucretia's cruelty. He bonds with Eleanor only when out of Lucretia's sight. Whenever Eleanor feels pain, she gains comfort from an old dog and from writing about her misery in "a book of secret thoughts."

Hawkes taps into the deep well of nineteenth-century pathos when he recounts the aunt's glaring eyes, snapping voice and rejecting attitude. The wicked stepmother motif, familiar from the peasant fairy tales of the brothers Grimm to middle class narratives, makes her plight almost archetypal. Eleanor persists in her quest for affection despite the rebuffs: "the frightened waif went to attack the very formidable castle in which abode the icy heart of her bitter, unforgiving Aunt."[140]

Lucretia converts because she snoops in Eleanor's book

and realizes the truth of accusations the child has penned against her. Like Augustine's biblical text, the journal forces her to confront her incivilities. Other tales told of people who also kept thought journals, most notably Mark Twain's *Pudd'nhead Wilson*, but they were not invested with this special power to change character. Nineteenth-century readers could parallel Eleanor's improving words with those of another fictional orphan. Robert Browning's sunny Pippa, "that ragged little girl," walks through a northern Italian town on her one day off from the silk mill. She sings a song that transforms murderers and adulterers when they accidentally hear it, a refrain that Hawkes quoted to cheer people up during the war-anxious 1940s.[141] *Wanted a Mother* ends when Lucretia fully accepts little Eleanor, makes new clothes for her, willingly cooks meals and invites Eleanor's friend from the poor-farm to become a live-in hired man. Thanks to the child's writings, Lucretia learns "to bottle up sunshine in her heart."[142]

Such nearly magical conversions caused by writing reinforced memories from Hawkes' own life. When first blinded, he suffered depression. His mother read Browning to him to allay his misery and, at Perkins, the books and lecturers opened up a world in which he had some purpose. Redemption by the word, a staple of Christian and particularly Protestant doctrine, worked for him at the personal level, saving him from isolation and dependency.

Similar renovations occur in *Doctor Thinkright*. Here the kindly old physician lives up to his allegorical name. With perception, Thinkright restores wounded hearts. The hypochondriac woman learns to put others first when the doctor adopts a spunky orphan and demands that she care for him. The stony heart of the miser Caleb Flint melts when the doctor maneuvers him into buying Christmas gifts for poor children and playing Santa Claus. The hypocritical factory owner, who chides the doctor for not attending church, reforms after being forced to visit people he has wronged—a poor washerwoman whose husband he had fired, a clerk who embezzled money to pay a high mortgage, a former worker in his unsanitary mill now dying of tuberculosis. In a strong subplot, the doctor becomes part of the lives of the town prostitute and a young alcoholic, whose lives also improve.

A mere synopsis cannot convey the meliorist assumptions of *Doctor Thinkright*. The title character may recall the dedicated healer who delivered Hawkes in Goshen so many winters before. Alternately, he may look back to Hawkes' great great grandfather, Dr. Enos Smith (c. 1771—1856). A graduate of Dartmouth, he practiced medicine in Ashfield at the start of the nineteenth century, but was remembered, in Hawkes' words, as a person "of whom so many witty stories are told."[143] Dr. Smith's houses had clocks, just as Dr. Thinkright's home.

However, the optimistic Thinkright most closely resembles Hawkes himself. The doctor has had "a life full of sorry" (15), although in public "His face was like the great sun showing above the eastern hilltops" (29). Similarly, "He had always loved children" (30), carries a cane, and has a brother in California (Hawkes' kid brother Ernest had earned a Ph.D. at the University of Pennsylvania in 1915 and, after various teaching jobs, moved in 1919 to Glendale,

California). In a maxim especially touching to those who know Hawkes' history, Dr. Thinkright says, "We are all blind to our own sins. It is only other people who see them" (25). The medical theory of counter-irritants that pained Hawkes as a newly blind teenager may resurface when the matron of the orphan asylum balks at releasing a mischievous lad to the complaining spinster: "Children are a sort of bombshell in a home where they are strangers" (33); Dr. Thinkright agrees, insisting, "You see, the child is medicine. It is to be a heroic dose, something so much worse than her ailments that she will forget them" (34). Perhaps remembering Dr. Smith's antic reputation, Hawkes lightens the seriousness of such shock therapy when little Tommy humorously dresses as a physician in oversize clothes and takes Edith's temperature with a "thermonica" (59-62). Like Eleanor, the old therapist himself derives comfort from "a little notebook" (16) of wise words.

Each conversion he effects alleviates "heart trouble." As he explains to pale Edith Grayson, the imaginary invalid who speaks with a "colorless voice" (19),

> The trouble with your heart is that it is supremely selfish. That is what makes you ill. You have not generosity enough to think of others, so you always and forever think of yourself. (23)

Sometimes the mechanism for redemption is a child like Tommy. George Eliot already had mapped out that plot in *Silas Marner* (1861), where the misanthropic weaver Silas takes in the foundling Eppie and learns love from her. Other times, as with greedy Flint, Hawkes uses the precedents of Augustine's twin life-altering encounters with a book and with disembodied voices. Hawkes has Flint speak like a second Scrooge from Dickens' *Christmas Carol*: "I don't ever think of anyone specially on Christmas, and no one thinks of me. It's mostly humbug" (75). Flint reads Dickens' novel and a current newspaper report of his generosity, explodes with anger, but then overhears a conversation among strangers that accuses him of stinginess. These assaults on his self-possession lead him to gaze into a mirror and, renewed, practice saying "Merry Christmas."

The *Master of Millshaven* revisits the subject of reformation from dark selfishness to positive social awareness. It follows in the great tradition of social protest novels that had piqued middle class consciences since Charles Dickens' *Hard Times* (1854), a fictionalized account of an unsuccessful strike by cotton workers in the Lancashire town of Preston. Dickens defended both the established owners and some of the protestors, vaguely suggesting better days would come when the human heart developed. Rebecca Harding Davis' grim report of *Life in the Iron-Mills* (1861) eliminated Dickens' poetic distance by presenting cold, hungry, tired, and doomed workers who would never escape their muddy, sooty lives. Her realism reappeared in Jacob A. Riis' *How the Other Half Lives* (1890), a series of essays and photographs that forced readers to see what poverty looked like in slums seldom visited by respectable audiences. Other books like Lincoln Steffens' *The Shame of Cities* (1904), which detailed corrupt city governments that neglected the needy while catering to the wealthy, and Upton Sinclair's *The Jungle* (1904), which exposed the horrendous Chicago meat packing factories, also lay behind Hawkes' narrative.

Published in 1936, with the Depression apparently confirming earlier fears that exploitative capitalism would collapse, Hawkes' novel eased anxieties by being set in the 1890s. The "master," a middle-aged man, was, like Lucretia in *Wanted a Mother*, once disappointed in love. He now pursues wealth at all costs, controlling mills, the police, banks, and newspapers in the three-tiered town of Millshaven. There, factories and poor workers mingle on the flats, the clerks and tradespeople live on the next higher terrace and the "aristocrats" on the top. The master's daughter, Evelyn, his foil, denies genetics and generously works with the poor. The male hero and Evelyn's love interest sounds like a younger and more practical Dr. Thinkright. Cullen Hayward, blinded accidentally by a friend, loves the country (as did William Cullen Bryant, Hawkes' Cummington predecessor and distant relative through William Clark Smith), but he becomes a lawyer in the city. There he combats a heartless plan by the master to further demean the lowly—building another factory on a playground.

Uncle Billy completes the quartet of conversion novels. Its title character, "The Curious Cobbler," had lost his wife and child two decades before in the west but now treats an entire eastern town as his family. He mends both soles and souls, having "discovered the appalling fact that every one is lonesome" (12). His good works benefit a wide range of victims. He donates money anonymously so the grade school teacher, Miss Browning, can pay for an operation to cure her young brother's deformed leg; he instills conscience into three mischievous students who had disrupted her class so they heartily apologize to her and become friends to her "cripple brother" (24); he reforms a gang of juvenile delinquents "from Tinpot Alley" (53) by inviting them to dinner and to a football game; he educates the cynical businessman Hiram Pepper so he can appreciate nature and generosity.

Uncle Billy draws strength to perform these charitable deeds from the same beliefs as those that motivated Hawkes. First, both accepted Wordsworth's idea that the child is closest to perfection and loses vital links to the universe as it ages. Pondering a pair of "battered little shoes," Uncle Billy muses: "They are miles nearer Heaven than we grownups. Sometimes, I think, the longer we live the further off we get from heaven" (15). Most adults must consciously recall this fact if they are to function altruistically. The miserly, no nonsense Hiram almost accidentally reads one of Thomas Hood's poems ("I'm farther off from heaven than when I was a boy") and begins to renew himself. (Significantly, like miser Flint in *Doctor Thinkright*, Hiram next "caught a glance of his face in a looking glass," an obvious but efficient way for readers to feel that the words have forced him to confront his true identity. [97-98].) The first token of Hiram's reformation is his approval of repairs to a tenement he owns. Later he reads the scene from Mark Twain's novel where Tom Sawyer observes his own funeral, an optimistic proof we can reverse the loss usually associated with living too long.

Another related idea that sustained both Uncle Billy and Hawkes was that injuries may be repaired. In the novel, children who can afford a hospital operation will escape from the life of a cripple. On the symbolic level, Hawkes has the Tinpot Alley gang, apparently hopeless victims of poor nurturing, destroy Uncle Billy's watermelon garden. The disturbing incident has powerful literary predecessors as old as Homer's *Odyssey*, where the well-tended garden of King Alcinous signified social order in the dream world of Phaeacia. Virgil's Old Man of Corcyra in *Eclogue 4*, although poor, also had a manicured patch, emblem of his complete happiness. To destroy anyone's plot implied sociopathic nihilism. Longus' second century CE pastoral romance *Daphnis and Chloe* has a disappointed suitor of the beautiful Chloe vandalize the garden of old Lamon, stepfather to Daphnis, whom Chloe loves to the exclusion of all others. Despite the symbolic gravity of such destruction, Hawkes employs the Rousseau-like notion that the four boys have been corrupted by society. Their devastation of Uncle Billy's melons does not end their connection to the public; rather it initiates their rehabilitation because, with proper guidance, all rogues may regain their earlier innocence. Their present state cries for improvement: one lives in a saloon, another has been forced to sell newspapers on the street since his earliest years. Uncle Billy invites them to dinner, forces them to confront their real identity when he calls each by his Christian rather than nickname, allows them to describe their upbringings, and sets in motion their personal confessions, contritions, and redemptions.

Along with valuing youth and trusting in recovery, Hawkes shared with Uncle Billy a third assurance about the value of tradition. The novel ends at a lavish Christmas dinner for 500 poor children. Domestic and public meals cement relationships. The original supper at Uncle Billy's prompted the rascals who ruined his garden to steal new clothing, slick down their hair, douse themselves with cologne, and try to observe proper etiquette at table; the Christmas feast brings together the town's abjectly poor (not just "the respectable poor" who do not "really need our love and pity" [159]), the three school cut-ups (who hand out invitations), Hiram (who pays for the event after "He had read Dickens' *Christmas Carol* the night before, and had been shocked to see how much like old Scrooge he had become" [148]), and Uncle Billy (who retells the Nativity story and moves everyone's emotions). The name of the banquet building, Foresters Hall, and the presence of a gigantic Christmas evergreen hint that all people can most easily mingle when they recall nature. The hall becomes a temenos, a sacred space, which for a brief moment exhibits the best reciprocal virtues of society purged of invidious distinctions based on wealth or physical condition. Hawkes' life-long involvement in other shared activities—sporting events, politics, and Hadley's 250th anniversary ceremonies—here finds its most potent symbol.

The four novels offered a uniquely American solution to social and spiritual problems investigated in England by E. M. Forster. *Howard's End* (1910) looked at the three separate worlds of the materialistic, hard-working bourgeoisie, the leisured and idealistic families supported by others' labor, and the striving children of country peasants, vainly trying to survive in harsh cities. Forster's epigrammatic solution, "only connect," sensitively implies that each class may live on if it links prose and passion, body and soul, manufacturing and culture. Hawkes added a vital dimension, that of nature, and foresaw a future good conclusion to the struggle of social contraries that wastes energy and distracts people from bonding.

Hawkes and Nightwatchman

Hawkes' Reputation

For such a multi-talented celebrity, one mystery remains: why is Hawkes not still famous in our day? He suffered so many cruel subtractions in his early years that he seemed doomed to a life of tragic passivity, but he displayed the strength that inspires legends and triumphed. In 1927 an essay on his life joined others profiling "the World's Famous Men" like Thomas Henry Huxley, "Darwin's bulldog," Leo Tolstoy and Oscar Wilde.[144] By 1931, everyone agreed that "Clarence Hawkes probably is the most widely known native of the Hampshire hills in the field of literature."[145] Fraternal groups like the Sons of Union Veterans of the Civil War,[146] the Knights of Pythias,[147] and the Lions Club [148] regularly honored him. Robert Bartlett's *They Dared to Live* profiled him along with Helen Keller, Oliver Wendell Holmes, and Charles Steinmetz as a model of courageous achievement.[149] As one modern website explaining the 1920s puts it,

"Clarence Hawkes was a famous author then."[150]

Yet folk memory has separated Hawkes from the company of penniless Ben Franklin, rural Abe Lincoln, and isolated Helen Keller. Today we must work to recapture his larger-than-life status in his own era. Fads govern fame, of course: Mrs. Felicia Hemans was once thought to be a better poet than Shelley. Most Pulitzer Prize, Nobel Prize, and best selling writers fall quickly off the public's radar. The passing of Hawkes' reputation may be caused by several modernist habits. First, accessible poetry like his no longer impresses audiences. Hawkes' predictable rhymes, sentiment, and generality contradict current expectations of idiosyncratic verse forms, sophisticated irony, private symbolism, anguished confession, and discontinuity. Next, the representation of nature has become a multi-media industry that requires high definition films of animal migrations, exciting memoirs of exotic field study, impressionistic hymns to the outdoors, glossy children's booklets, and cartoon simplifications. This deluge of offerings, enhanced by stereophonic sound or celebrity voice-overs, understandably edges out Hawkes' well-made but one-sense reports. In addition, the quest for social justice that Hawkes felt would best be served by a conservative Yankee has been questioned as nativism or paternalism. His admiration for Calvin Coolidge or Warren G. Harding or even Herbert Hoover now sounds as if dated wishful thinking rather than understanding governed his preferences.

Although early twenty-first century taste has altered, we still can admire the qualities that allowed Hawkes to produce so much. His Perseverance, Pluck, and Patience, far from being abstractions, actualized his tremendous potentials for creativity and kindness. His feats of memory alone would impress anyone, especially in his day when even trained orators and preachers spoke with the help of texts. The poetry urged conservation of time-tested virtues and decent sensations. The nature books fleshed out with imagination the facts of non-human life that, properly understood, would benefit all creatures. His novels offered an alternative view of a world temporarily damaged by rapacious businesses, selfish hearts, and gloating materialism.

Sylvia Niedbala
Woicekoski

Theodore C.
McQueston

How Hadley Now Remembers Hawkes

Current Hadley residents preserve a positive collective memory that confirms his national reputation for these achievements. As Theodore (Ted) C. McQueston (1919–) put it, "Everyone knew him." McQueston's older sisters, Dorothy and Ruth, proofread Hawkes' typewritten manuscripts. Once, young Ted strolled over to observe them at work. His mother telephoned and Hawkes said that the boy wasn't there. "Yes, he is," shouted one of the girls. Hawkes gave him a nickel for being so quiet; Ted's mother, however, gave him "heck" for disappearing.

When older, McQueston drove Hawkes to Goshen and marveled at the sightless man's ability to say where they were on the journey.[151] McQueston relished the many books

CHARLES J NIEDBALA

Hawkes gave his family. One favorite, *Patches, a Wyoming Cow Pony*, startled him with its accurate depiction of equine behavior. The McQuestons owned horses and he felt "it was amazing he could describe so many animals without having seen them." Hawkes reciprocated the respect: when McQueston's father died, he walked over to pay his respects.[152]

Many judged him as a fair employer. Like the McQueston sisters, Hawkes' former neighbor Eleanor Miller worked for him when she was a teenager: "There were a few of us girls who would read to him. He would type his copy and after he was done with a page we would read it back to him to see if there were any errors."[153] Walter C. Wanczyk, Jr. (1948—), whose father did work for Hawkes, recalls hearing how he paid generously for services. Careful of finances, though, he would accept change from neighborhood children in a metal can so that the clink of coins allowed him to add up the amount.

ANTONIA S. DECJAKOBEK

Sylvia Niedbala Woicekoski (1925—) also did errands for Hawkes. Living on Russell Street (then 46, now 100) at the corner of West, she easily crossed the road to Shipman's store, retrieved his mail from the small post office, and delivered it to a wooden box outside the home. For this vital task, each Saturday morning at 10 he gave her $ .35, a liberal sum. And she would bring bouquets of violets to Bessie, who usually answered the door. Whether or not Bessie realized it, the flowers had been picked in the Hawkes' own back yard. The sight of brother Enos, tall and skinny as opposed to Hawkes' sturdy frame, made her think of Ichabod Crane, the lanky schoolteacher in Washington Irving's "The Headless Horseman of Sleepy Hollow." He was not the only live-in helper. Once a friendly housekeeper invited her to the second floor room at the rear of the modest home. Only two regrets slightly diminish Mrs. Woicekoski's pleasant recollections: she lent friends three books Hawkes had inscribed to her and never got them back; in addition, James P. Reed, the popular Principal of Hopkins Academy for an unprecedented thirty-five years (1914-1949), a frequent visitor, had a large English setter that he called Sylvia.[154]

Charles (Charlie) J. Niedbala (1936-), Sylvia's brother, mowed Hawkes' lawn in exchange for $ .20. The pay seemed fair; another customer, whose lawns took two days to cut, paid $1.00. He recalls a few apple trees in the back yard, the ones that Mr. Burke, the Hopkins industrial arts teacher, had once mobilized his students "like the Assyrians of old" to pick and pack. Otherwise, Hawkes' Baldwin and Greening apples would have fallen uselessly to the ground.[155] Niedbala vividly remembers how Hawkes would rock on his front porch and hum to himself.

Author of many musical compositions, ranging from (the embarrassingly titled) "A Pickininny Lullaby" to anthems for Hopkins Academy, he may have been rehearsing one of them. At several times, assistants appeared at the house. One, brother Enos, was taller than the sturdy Hawkes, perhaps a reminder of uncle William at the age of 28, when his Civil War enlistment paper recorded him as being "5'10 ½."[156] A second aide named Mable, whom Niedbala thinks may have been from Springfield, also worked in the house.[157]

Hawkes' lack of prejudice often surfaced. Antonia (Ann) S. Dec Jakobek (1918-), who lived next to him on West Street, was sister to Stanley Dec, whose premature death after a fall from a tobacco rack inspired Hawkes' touching eulogy, "My Little Polish Neighbor." In it he sensitively confronted the mystery of unmerited loss. He consoled himself and his townspeople by using an agricultural metaphor: "I like to think of Stanley as just transplanted to some more beautiful garden of 'God's Little Ones.' A garden where the sun is brighter and everything more beautiful than it can ever be here." Admittedly, he used the familiar Victorian tropes of mourning a child, of imagining a pastoral afterlife, and of hoping that the life of an innocent person never ends. But they spoke to his generation and, perhaps, to the majority of readers in every era who do not look for modernist sophistication when they wish to express deep emotion. Most important, the elegy proves Hawkes did not share the unfortunate suspicion of Poles that clouded some residents' judgment. Ms. Jakobek reinforces the picture of a town that willingly took care of Hawkes, seconding the fact that Principal Reed was one of many chauffeurs who sought out his company for road trips.[158]

Mrs. Jakobek still remembers how he enjoyed the youngsters who would pass by as he and Bessie sat on their porch, proof of the epitaph on their Old Hadley Cemetery marker: "They loved nature and little children." Amazingly, he recognized many of them by their voices. Bessie often made little paper dolls that youngsters could fasten to their shirts.[159]

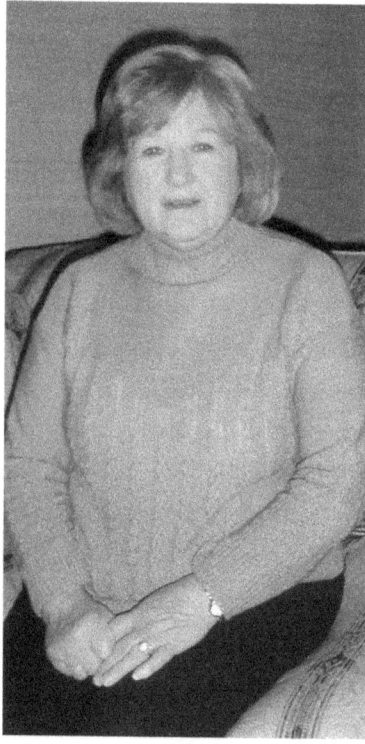

DIANE KOZERA BAJ

Diane Kozera Baj (1942-), who lived at 61 West Street,

seconds the affection the Hawkes had for children. She, her friend Lorainne Michalowski Zieminski, who lived in the Bell house, and their playmates would notice the couple on their porch with their small dog. Because there was a sidewalk, kids could ride past their house and yell greetings to them. Often, he would come out and reply to their polite, "Good morning, Mr. Hawkes" with "And who are you?" He engaged youngsters, once they identified themselves, with caring questions ("Is your father still farming?"). The couple "enjoyed us, yelling and screaming, as we kicked the ball on the common." At Halloween, youngsters received a warm welcome at his door when they threatened trick or treat. He would affectionately ask them to describe their costumes; then Bessie handed out the candy. Although he differed from other men in that he wore cloudy glasses, had a cane ("He was pretty good with it"), and often rocked or strolled back and forth on the front porch, Ms. Baj knew that people "had high respect for this man." Poetry naturally meant little to her as a child, but her parents characterized him as a "gentle, kind man." He had the ability to converse with all visitors because he was "always interested in everything in the community." He especially enjoyed the sounds of music issuing from carnivals on the common.

Miriam (Midge) R. Pratt (1915—2009) amusingly recalls her high school days in the old Hopkins Academy. Class of 1934, she would, like generations of pupils, gaze out of the building's large windows during class. She recognized Hawkes when he walked to the edge of Route 9, waiting for someone to guide him safely between the few cars and trolleys, to Shipman's store on the north side.[160]

Miriam (Midge) R. Pratt

Hawkes repaid as best he could the generosity of his neighbors. Michael J. Lesko, Jr. (1933—) lived at 107 Russell Street, the south side of Route 9, east of the current (2009) Neidbala store, just a few yards from the Hawkes' residence. When Lesko's father built their house in 1924, Hawkes obeyed the local custom of presenting a house-warming gift. The black wooden rocker still proudly sits in the present Lesko home, surrounded by other mementoes of Hadley history: photos of the great flood of 1936, a tobacco rack like that which crushed little Stanley Dec, the now burned wooden covered bridge and, nearby, a print brought from Poland by Lesko senior showing a girl like the heroine of "Little Sister."

Lesko used to play on the Hawkes' porch and remembers looking through the floor-to-ceiling windows as Hawkes typed. "He always seemed to be working there," confirmation of Hawkes' claim to have worn out a half score of typewriters. Sometimes an assistant (perhaps brother Enos or William) helped. The neighborhood children may have been polite while he worked, but, being young, they rummaged through Hawkes' scrap heap. "I never saw anyone drink so much tea," Lesko mused, confessing that he and his pals took used tea bags. Small cloth pouches, they weren't practical for further use except to carry a few marbles.[161]

CLARENCE HAWKES READING BRAILLE

The Lesson of Clarence Hawkes

An old myth claims that gods perish when their last worshipper passes on. At present, only long-time residents think much about Clarence Hawkes. Hopkins Academy students probably have little idea who wrote their school songs, "Hail! Hopkins, Hail!" and "Hopkins, My Hopkins."[162] His writings appear on used book dealers' lists, but their dynamic benevolence no longer reaches readers. Today's audiences, already drowning with urgent solicitations to earn more money, lose weight, whiten teeth, purchase electronic gadgets, and take a cruise, would profit from his example. Ironically, the industrious creator turned out to be the true hedonist: he overcame pain and discovered a complete, satisfying life. "[T]he blind (but never sightless) man"[163] took justifiable pride in his attainments and left his later admirers with a heroic American image:

> Every inch of the way I have fought. No miner delving for gold in the frozen Arctic, with the thermometer at sixty below zero, and the earth frozen for God only knows how far down, has ever sweated and struggled the weary march, loaded down by his heavy knapsack and gun, with the mud halfway to his knees, has ever had to fight as I. My success, what little I have gained, has been literally dug out of the solid rock of adversity, with naked, bleeding fingers.[164]

Clarence Hawkes Genealogy

```
                    Dr. Enos Smith ─────── Hannah Woods Ware
                    (16 Jan 1770-11 Oct 1856)  (31 Jan 1775-13 Dec 1839)
                         (married 25 Jan 1797 at Conway)
                         (both buried Granby West Cemetery)

William Hawkes ── Abigail Nabby Marsh
(c. 1774-6 June    (c. 1772-17 Nov 1842)
1817. Deerfld)     (Conway-Worcester)
    (married 19 Nov 1794 at Deerfield)


   William Hawkes      Almira Smith      Josiah Gurney     Emily Bates Meritt Jenkins
   (22 Sep 1807-       (15 May 1809-     (15 July 1806-    (c. 1820-12 Dec 1877)
   29 Mar 65)          8 Aug 1847)       4 Nov 1886)
   (Deerfield-Hamp)    (Ashfield-Ashfield)   (married after 26 Aug 1843 at Ashfield)
        (married 4 July 1832 at Ashfield)   (both buried Ashfield Spruce Corner Cem)


   William Smith Hawkes    Enos Smith Hawkes    Edlah Betsey Olive Bates Gurney
   (1 Mar 1834-            (22 Sep 1840-        (20 Jan 49-13 Oct 1899)
   18 May 1922)            18 Aug 1894)
   (Ashfield-Hadley)       (Ashfield-Hadley)    (Ashfield-Northampton)
   (married 3d wife            (married 29 Nov 1866 at Ashfield)
   Alma Lillia Bates          (both buried Ashfield Hill Cemetery)
   28 June 1876 at
   Buckland. Alma
   d. 14 Oct 1900)


   Clarence      Alice Edith      Enos Raymond    Arthur Josiah    Ernest William
   E[nos]        (22 July 1872-   (2 April 1875-  (21 June 1877    (19 July 1881-
   (16 Dec       30 Ap 1897)      ? )             -2 Feb 1941)     13 Mar 1957)
   1869-19       (Goshen-         (Goshen- ? )    (Goshen-         (Ashfield- Los
   Jan 1954)     Hadley)                          Gardner)         Angeles)
   (Goshen-      (buried
   N'hampton)    Ashfield Hill
   (buried       Cemetery)
   Old Hadley)
```

Edlah Gurney Hawkes
Clarence's Mother

Josiah Gurney——**Olive Torrey**   **Newton Bates**—**Betsey** (family?)
(29 April 1785-   (c. 1783-13
4 Jan 1862)   Jan 1861)
    (married 4 Dec 1805)

      Josiah Gurney————————————**Emily Bates Meritt Jenkins**
      (15 July 1806-4 Nov 1886)   (c. 1820-12 Dec 1877)
      (? -Savoy)   (Westford, VT-Ashfield)
          (married after 26 Aug 1843 at Ashfield)
          (both buried Ashfield Spruce Corner Cemetery)

Enos Smith Hawkes————————**Edlah Betsey Olive Bates Gurney**
(22 Sep 1840-18 Aug 1894)   (20 Jan 1849-13 Oct 1899)
(Ashfield-Hadley)   (Ashfield-Northampton)
    (married 29 Nov 1866 at Ashfield)
    (both buried Ashfield Hill Cemetery)

                Samuel Reuben Bell——**Sarah A. Wilder**
                (c. 22 Oct 1839-5 Feb   (c. 1843-22 Oct
                1920)   1913)
                (Hadley-Hadley)   (Mosco, NY-Hadley)
                  (both buried Old Hadley Cemetery)

            Clarence E[nos] Hawkes————**Bessie Wilder Bell**
            (16 Dec 1869-19 Jan 1954)   (27 Aug 1869-24 Jan 1958)
            (Goshen-Northampton)   (Hadley-Hadley)
              (married 30 October 1899 at Hadley)
              (both buried Old Hadley Cemetery)

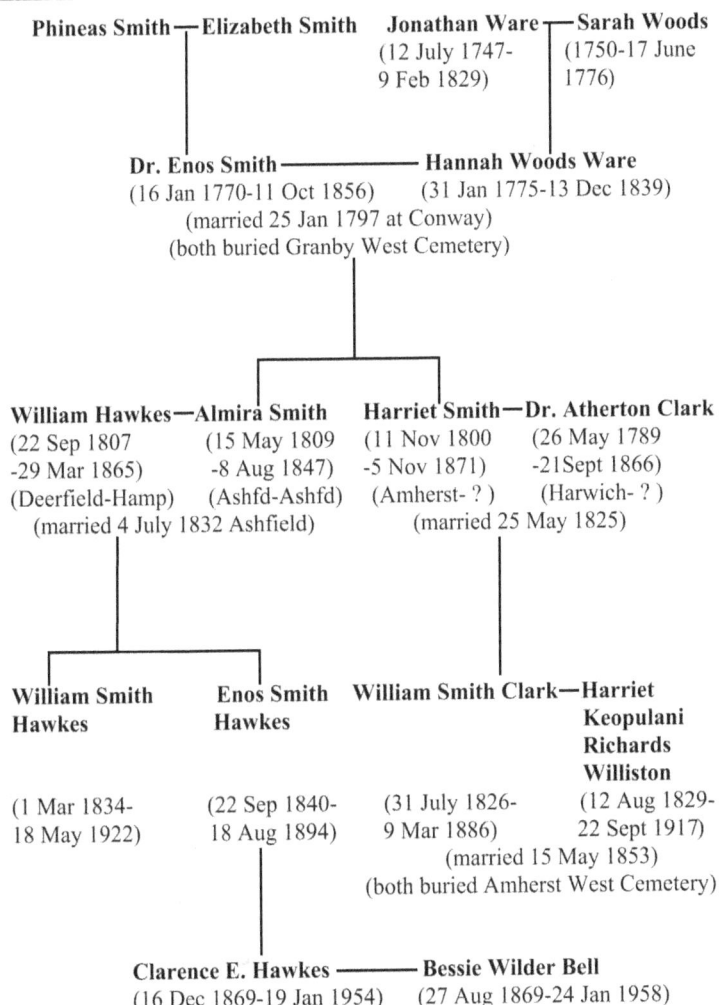

APPENDIX: Nature Writing Before Hawkes

To simplify a particularly rich half-century, say the mid-1800s through the first decades of the twentieth century, one can point to two ways that authors whom Hawkes respected addressed their audiences on the topic of nature. Following Romantics like William Wordsworth, some tried to replace conventional Christian theology with a heightened reverence for the universe. Salvation from anxiety and instruction for action, they claimed, could come from the non-human world. As Wordsworth said in the passage already quoted, nature "May teach you more of man /, .../ Than all the sages can." Fashionable writers espoused the hope that retreat to physical nature might open in them a vision of infinity. As Ralph Waldo Emerson famously said,

> Standing on the bare ground,--my head bathed in the blithe air, and uplifted into infinite space,--all mean egotism vanishes. I become a transparent eyeball; I am nothing; I see all; the currents of the Universal Being circulate through me; I am part and parcel of God.[165]

Hawkes and his fictional speakers often sounded like their disciples. One pair of his affirmations may stand for many others. He begins *Notes of A Naturalist*, "There's a balm in Mother Nature, / For earth's heartache and its pain," and then asserts, "The seeing eye is not an acquired faculty, but a gift of God vouchsafed only to the child of Nature.... This special vision is a spiritual quality."[166]

Less mystically, the idyllic poems of William Cullen Bryant assumed people who communed with seasons, creatures and plants could access some indwelling spirit. His most famous poem, "Thanatopsis," began typically, "To him who in the love of Nature holds / Communion with her visible forms, she speaks / A various language."[167] All of us, the theory ran, can appreciate the enchanting exterior of nature and perhaps her inner significances.

Unfortunately, most authors of nature books for children, like Hawkes' contemporaries Beatrix Potter (1866-1943) and Thornton Burgess (1874-1965), trivialized and sentimentalized these powerful beliefs. Their beasts were humorists, innocuous playthings that frolicked and chattered in a world with little transcendent meaning. In Burgess' *The Adventures of Jerry Muskrat* (1914), the inhabitants of the Smiling Pool and Laughing Brook have "conventions" at the Big Rock, all speaking the same language and agreeing to work together to avoid traps set by the farmer's boy. Most unsettling, the illustrator dressed them in surreal, English gentleman costumes—top hats, frock coats and umbrella or walking stick for Grandfather Frog, Ol' Mistah Buzzard and Spotty the Turtle. These Disneyfied creatures reassured their audiences that the world is human-centered, benign, sexless and predictable. They demonstrated how in some children's literature the Romantic passion for a sublime union with the non-us fizzled into saccharine amusement at creatures that existed only in patronizing imaginations. The recognizable antics of the wee beings proclaimed that people already know everything about the natural world—it mirrors human society—and thus they need not explore it.

Still, animal fantasies, like those of Hawkes' near age

mate Rudyard Kipling (1865-1936), sold. *The Jungle Books* (1894 and 1895), with Mowgli, the human orphan educated in India by wolves, a bear and a panther, his surrogate parents, plus *Just So Stories* (1902), announced that any author wishing to compete might have to refashion the public's expectations, however unreal, in nature tales. The implications of Kipling's first books—that humans can always rule animals and, by extension, that "civilized" countries should control less "advanced" ones—might have interested Hawkes. However, the etiological particulars of just so stories like "How the Whale got his Throat," "How the Camel got his Hump" or "How the Leopard got his Spots" bore no relation to evolution and were communicated in a stilted Anglo-Indian dialect which, while appealing to many, may have bothered Hawkes. In addition, Kipling did not imply that all life interconnects, favoring, as he did, certain creatures over others and rejoicing at the near extinction of those he disliked (such as the wild dogs stung to death by bees in *The Jungle Book*).

Early in Hawkes' career, he occasionally sounded like his British predecessor when he told how the turtle got his shell and how the fox got his tail. *The New York Times* reviewer of Hawkes' 1917 *Wood and Water Friends* linked the two writers: "Some of the tales are after the manner of '*The Jungle Books*'; that is, the birds and animals are themselves the chief 'characters,' talking among themselves."[168] But Hawkes emphasized their difference when he titled his 1935 autobiography *The Light That Did Not Fail*, a refutation of Kipling's 1890 *The Light That Failed*.

Other authors marshaled against this benign flock

a savage horde of Darwinian animals. The nineteenth-century Naturalists accepted Isaac Newton's proof that the cosmos ran according to mechanical laws independent of human desires. Edgar Allan Poe's grotesque "The Murders in the Rue Morgue" (1841) had described a run-away "Ourang-Outang" who killed two elderly women in Paris mainly because he could. Just before Hawkes' birth, Charles Darwin's *Origin of Species* (1859) showed how an inexorable biological imperative killed off inefficient species and preserved only those that out muscled their competitors. Ouida's melodramatic *A Dog of Flanders* (1872) presented a model trio that initially would have captivated Hawkes—a young and artistically talented orphan, his ailing grandfather and their loyal dog—but she allowed poverty in the Dutch countryside to crush the life out of them with no hint of cosmic explanation, social betterment, or moral enlightenment. (After the grandfather dies, the boy and dog freeze to death in an Antwerp church.) Even Karl Marx's belief that all group life demanded a fight between the rich and the poor added to impatience with the nursery school narratives about beasts.

Some thinkers applied Alfred Lord Tennyson's 1850 catch phrase "Nature, red in tooth and claw"[169] to people well as to brutes. Occasionally, as in Herman Melville's *Moby Dick* (1851), a mammal like the great white whale symbolized the world's cryptic evil and infected men like Captain Ahab. Upton Sinclair responded to the degradation of men and cattle in Chicago with his fitly titled novel *The Jungle* (1906). Unlike those living in, say, Kipling's tropical forest, everyone in the city did not gather in parliamentary-style pack meetings, did not obey a protective "law of the jungle," did not speak a universal language and did not furnish a "stranger's hunting call" to protect someone like Mowgli who roved beyond his normal territory. Theodore Dreiser, too, assumed the only lesson one learns from creatures is that you brawl or die. His novel *The Financier* (1912) begins as a young man watches a squid and a lobster fight to their deaths. This lethal spectacle energized him to become a millionaire, seizing control of Chicago's rail system and exploiting women. Arthur Conan Doyle likewise separated human from savage in his popular Sherlock Holmes adventures. "The Speckled Band" (1892) pictured a villainous physician returned from India, who atavistically associated with gypsies and wild beasts while he used a trained snake to murder his ward. *The Hound of the Baskervilles* (1902) featured a seemingly immortal and ferocious "gigantic hound" that terrified Dartmoor, similarly requiring Holmes' ingenuity to uncover the twisted human criminal.

Two final popular authors pondered the possibly destructive links between human and animal life, contradicting the impartial vision Hawkes was to communicate. H. G. Wells' disturbing *The Island of Doctor Moreau* (1896) wondered whether science might turn beasts into semi-human beings. Its conclusion: animals may temporarily resemble us, but they will inevitably obey their lust for blood and revert to feral predators. The most relentless exponent of the savage wildlife theory, Jack London (1876—1916), assumed that domesticated animals will quickly revert to savagery (*The Call of the Wild*, 1903) and that humans do not differ from beasts: he names one main character, who quoted Shakespeare but did not hesitate to kill, Wolf Larson (*The Sea Wolf*, 1904).

DANISH TRANSLATION OF *Wanted a Mother*. SUBTITLED "THE HISTORY OF A LITTLE HEROINE." TRANSLATED BY ERNA DAMGAARD (COPENHAGEN: CHR. ERICHSENS, 1923). NOTE SPELLING OF HAWKES' NAME.

CLARENCE HAWKES BIBLIOGRAPHY

Pebbles and Shells. Illustrations by Elbridge Kingsley. Northampton, MA: Picturesque Publishing, 1895.

Three Little Folks: Verses for Children. Northampton, MA: Picturesque Publications, 1896.

Idyls of Old New England. Illustrations by R. Lionel De Lisser and Bessie W. Bell Hawkes. Northampton, MA: Picturesque Publishing, 1897.

Songs for Columbia's Heroes: War Poems for 1898. Illustrations from Photographs by R. Lionel De Lisser and from Paintings and Drawings by Other Artists. Springfield, MA: New England Publishing, 1898.

The Hope of the World and Other Poems. Illustrations by R. Lionel De Lisser, Clifton Johnson, and Bessie W. Bell Hawkes. Springfield, MA: New England Publishing, 1900.

The Hope of the World and Other Poems. Illustrated by R. Lionel De Lisser, Clifton Johnson, Bessie W. Bell Hawkes. 2nd edition. Springfield, MA: New England Publishing, 1900.

Master Frisky. New York, T.Y. Crowell, 1902.

The Little Foresters: A Story of Field and Woods. New York: T. Y. Crowell, 1903.

Stories of the Good Green Wood. Illustrated by Charles Copeland. New York: Thomas Y. Crowell, 1904.

Shaggycoat; the Biography of a Beaver. Illustrations by Charles Copeland. Philadelphia, G. W. Jacobs, 1906.

The Trail to the Woods. New York, Cincinnati: American Book, 1907.

Tenants of the Trees. Boston, L. C. Page, 1907.

The Little Water-folks; Stories of Lake and River. New York: T. Y. Crowell, 1907.

Black Bruin; the Biography of a Bear. Philadelphia: George W. Jacobs, 1908

Shovelhorns, the Biography of a Moose. Philadelphia, G. W. Jacobs, 1909.

A Wilderness Dog; the Biography of a Gray Wolf. Philadelphia: G. W. Jacobs, 1910.

King of the Thundering Herd: the Biography of an American Bison. Illustrated by Charles Copeland. Philadelphia: G.W. Jacobs, 1911.

Piebald: King of Bronchos: the Biography of a Wild Horse. Illustrated by Charles Copeland. Philadelphia: George W. Jacobs, 1912.

The Boy Woodcrafter. Chicago: F. G. Browne, 1913.

Field and Forest Friends; a Boy's World and How He Discovered It. Illustrated by Charles Copeland. Chicago: F.G. Browne, 1913.

King of the Flying Sledge; the Biography of a Reindeer. New York: H. Holt, 1915.

Hitting the Dark Trail; Starshine through Thirty Years of Night. Illustrated by Charles Copeland and from Photographs. New York: H. Holt and Company, 1915.

Wood and Water Friends. Illustrated by Charles Copeland. New York: Thomas Y. Crowell, 1917.

"Angela—A Love Story." *Book News Monthly.* Reprinted in *Current Opinion* 62 (April 1917): 281-82.

Trails to Woods and Waters. Illustrated by Charles Copeland. Philadelphia: G.W. Jacobs, 1920.

Tenants of the Trees. Illustrated by Louis Rhead. Philadelphia: G.W. Jacobs, 1921.

Wanted a Mother. Philadelphia: G. W. Jacobs, 1922.

Pep. The Story of a Brave Dog. Illustrated by William Van Dresser. Springfield, MA: Milton Bradley, 1922.

Shipmate. Evanston, IL: Signal Press [c. 1923].

The Way of the Wild: Stories of Field and Forest. Illustrated by Charles Copeland. Philadelphia, PA: G. W. Jacobs, 1923.

The White Czar; a Story of a Polar Bear. Illustrated by Charles Livingston Bull. Springfield, MA: Milton Bradley, 1923.

The Way of the Wild; Stories of Field and Forest. Philadelphia, G. W. Jacobs, 1923.

Dapples of the Circus: the Story of a Shetland Pony and a Boy. Illustrated by L. J. Bridgman. Boston: Lothrop, Lee & Shepard, 1923.

"Little Sister." *New England Homestead* 23 June 1923: 13. Reprinted in "Hadley Story by Clarence Hawkes," *Daily Hampshire Gazette* 7 August 1923, page 7, columns 4-5.

Silversheene, King of Sled Dogs. Illustrated by Charles Livingston Bull. Springfield, MA: Milton Bradley, 1924.

A Gentleman from France. An Airedale Hero. Illustrated by L. J. Bridgman. Boston: Lothrop & Shepard, 1924.

Wood and Water Friends. Philadelphia: Macrae Smith, 1925.

Pal o' Mine, King of the Turf. Illustrated by Charles Livingston Bull. Springfield, MA: Milton Bradley, 1925.

"Clarence Hawkes in Tribute to Dec Boy," *Daily Hampshire Gazette* 24 August 1926, page 4, column 1. Reprinted in Hawkes, *The Light That Did Not Fail*, 100-103.

Jungle Joe, Pride of the Circus; the Story of a Trick Elephant. Illustrated by L. J. Bridgman. Boston, Lothrop, Lee & Shepard, 1926.

Redcoat, the Phantom Fox. Illustrated by Charles Livingston Bull. Springfield, MA: Milton Bradley, 1927.

Patches, a Wyoming Cow Pony. Illustrated by Griswold Tyng. Springfield, MA: Milton Bradley Company, 1928.

Bing, the Story of a Small Dog's Love. Boston, Lothrop, Lee & Shepard, 1929.

Big Brother; the Story of a Trick Bear. Illustrated by Griswold Tyng. Springfield, MA: Milton Bradley, 1930.

Peter, the Story of Little Stoutheart. Boston, Lothrop, Lee & Shepard, 1931.

Doctor Thinkright. New York: Thomas Y. Crowell, 1934.

The Light That Did Not Fail. Boston: Chapman & Grimes, 1935.

Roany, The Horse who Smelled Smoke. Springfield, MA: Milton Bradley, 1935.

The Master of Millshaven. Boston : Chapman & Grimes, 1936.

Igloo Stories: Six Tales of Eskimo Land. Boston: Christopher, 1937.

Christmas All the Year. Boston: Christopher, 1938.

Notes of a Naturalist: Jottings in Field and Forest. Boston: Christopher, 1938.

Uncle Billy, the Curious Cobbler. Boston: Chapman & Grimes, 1939.

Holiday Hopes. Boston: Christopher, 1939.

My Country; the America I Knew. Boston: Chapman & Grimes, 1940

Dapples of the Circus. Boston: Lothrop, Lee & Shepard, 1923; New York: Platt & Munk, 1943.

Big Brother; the Story of a Trick Bear. Illustrated by Griswold Tyng. New York: Platt & Munk, 1943.

The Strange Adventures of Mr. Turtle. New York, Harbinger, 1944.

The Service Man's Friend. Boston: Chapman & Grimes, 1944.

NOTES

No historian works alone. In addition to the talented and supportive people gratefully but inadequately acknowledged in the notes below, several wonderful new friends helped this essay. Jane Babcock, Librarian, Goodwin Memorial Library, Hadley, enthusiastically supplied pictures and bibliographical facts. Carol Booker, Hadley genealogist, expertly researched complicated family relationships. Jessica Spanknable, Hadley Town Clerk, quickly found vital statistics for the Hawkes and Bell families. Elise Feeley, Reference Librarian, Forbes Library, Northampton, skillfully provided biographical data, especially Kelly's MA thesis. Vanessa Fish, Town Clerk of Goshen, graciously retrieved vital statistics about Clarence, Alice, Arthur, and Enos. Nancy Gray Garvin, Ashfield historian, deftly transported me to Clarence's school and family tombstones. Ruth Owen Jones, Amherst historian, engagingly shared her extensive understanding of William Smith Clark. Steve Mollison, Goshen Historical Commission, generously mailed an important notice from a town history. Stephanie Pasternak, Secretary to the Cummington Historical Commission, kindly forwarded an obscure but useful bibliographic reference. Norene Roberts, Chair of the Goshen Historical Society, courteously contributed valuable data about Hawkes' relation to her town. Because others contributed so much so efficiently, I can take credit only for errors.

ENDNOTES

1 Clarence Hawkes, *Hitting the Dark Trail; Starshine through Thirty Years of Night* (New York: H. Holt, 1915), 8.

2 "Clarence Hawkes in Tribute to Dec Boy," *Daily Hampshire Gazette* [Northampton, Massachusetts] 24 August 1926, page 4, column 1. Reprinted in Hawkes, *The Light That Did Not Fail*, 100-103.

3 Clarence Hawkes, "Little Sister. A Story of Connecticut Valley," *New England Homestead* 86-87 (23 June 1923): 13.

4 Clarence Hawkes, "Angela—A Love Story," *Current Opinion* 62 (April 1917): 281-82.

5 Clarence Hawkes, *My Country. The America I Knew* (Boston: Chapman & Grimes, 1940), 189.

6 "Clarence Hawkes is Included Among 35 Who 'Dared to Live,'" *Daily Hampshire Gazette* 10 November 1937, page 4, column 2.

7 Frank G. Jason, "Clarence Hawkes, Blind Poet, Celebrates 60 Years of Bringing Courage to Sightless," *The Boston Sunday Post* August 1943.

8 Clarence Hawkes, *Notes of A Naturalist. Jottings in Field and Forest* (Boston: Christopher, 1938), 46-47.

9 "Clarence Hawkes Speaks at Goshen," *Daily Hampshire Gazette* 15 August 1931, page 1, column 5.

10 Clarence Hawkes, *The Service Man's Friend* (Boston: Chapman & Grimes, 1944), 59.

11 The doctor who had to trudge uphill through the whiteout snow may have been Thomas Gilfillan. According to Stephen M. Howes, former Chair of the Cummington Historical Commission, the physician might have started from the house he occupied from 1856-c. 1874 on Cummington's Main Street—ironically, next door to the house later occupied from 1885-1892 by the Hawkes family. Gilfillan would have headed up the steep slope. Leaving behind his exhausted horse, the

determined doctor was helped by a "kind farmer" (perhaps Howes' own great grandfather Jesse Willcutt, who had lived since 1857 in the last house in Swift River on the route to the Hawkes' place). (Howes personal interview. 21 December 2006.) Hawkes describes the winter birth in *The Light That Did Not Fail* (Boston: Chapman & Grimes, 1935), 12-17. The kindly physician may reappear as Dr. Thinkright.

12   Illinois Civil War Records. *http://ilsos.gov/genealogy/CivilWarController*.

13   Hawkes, *Hitting the Dark Trail*, 17.

14   *The Light That Did Not Fail*, 30.

15   Date of award: Imogene Hawks Lane, *John Hawks. A Founder of Hadley, Massachusetts. After a Sojourn of Twenty-Four Year at Windsor, Connecticut. Thirteen Generations in America* (Baltimore: Gateway Press, 1989), 390; Amount of award: *List of Pensioners on the Roll, January 1, 1883*, 2 volumes (Baltimore: Genealogical Publishing, 1883; reprint 1970), 1. 354.

16   "Ashfield," *The* [Greenfield, Massachusetts] *Gazette and Courier* 23 June 1884.

17   "Chesterfield and Goshen," *County Atlas of Hampshire Massachusetts from Actual Surveys under the Direction of F. W. Beers* (New York: F. W. Beers, 1873), 14-15.

18   David Starr Bingham, *The Banister House. The Story of a House and the Families Associated with It* (Privately printed, 1999). In later years, citizens of Goshen decided to place a memorial on "the six-room cottage" and "the district school house where [Clarence] learned his three R's from 1877 to 1890" ("A Son of Goshen," *Daily Hampshire Gazette* 9 July 1931, page 8, column 2). That homage was accorded to nationally famous figures like the regicide judges who hid from Charles II's vengeful agents and Civil War General Joseph B. Hooker, both in Hadley. Today only the small Spruce Corner School survives, but a handsome bronze tablet honoring Hawkes glows high on the wall of Goshen's public library. It was placed originally in the community building on 2 August 1939. (Hawkes' note, *My Country*, 117.)

19 "We are happy to report," *Hampshire Gazette and Northampton Courier* 28 August 1883, page 2, column 8.

20 *The Light That Did Not Fail*, 80.

21 Lane, *John Hawks*, 390.

22 Anne Sabo Warner, *A Bicentennial History of Goshen, Massachusetts 1781-1980*, (Goshen: Historical Commission, 1980), 188, 373.

23 Hawkes, *Hitting the Dark Trail*, 14-15. Typically, Hawkes followed up the connection with Norton. The respected Harvard professor of Fine Arts, translator of Dante, and former editor of the *North American Review, Atlantic Monthly*, and *The Nation* convened famous lecture dinners in Augusts from 1879 until 1903 at the academy building. More than 140 speeches from luminaries like G. Stanley Hall, President of Clark University, Charles Dudley Warner, Editor of the *Hartford Courant*, author William Dean Howells, President Booker T. Washington of Tuskegee Institute, and James Russell Lowell enriched auditors' minds (Betty and Edward Gulick, *Charles Eliot Norton and the Sanderson Dinners, 1879-1903* [Ashfield: Ashfield Historical Society, 1990]). Norton's objections to American culture as puerile and crass in addition to his opposition to our empire building in Cuba may not have convinced the grown-up Hawkes, who was consistently patriotic, but the sharing of ideas appealed to him early and late. Norton praised Hawkes because his writing did not "minister to the popular taste for gossip" (Letter from Charles Eliot Norton to Hawkes, 6 April 1895, Hawkes Manuscript Letters 1. 6). Later Norton ordered two of Hawkes' books (Letter from Charles Eliot Norton to Hawkes, 20 October 1906, Hawkes Manuscript Letters 1. 66).

24 Clarence Hawkes, *Roany, The Horse who Smelled Smoke* (Springfield, MA: Milton Bradley, 1935), 74-76.

25 Clarence Hawkes, *Igloo Stories: Six Tales of Eskimo Land* (Boston: Christopher, 1937), 121-22.

26 Editors of Time-Life Books, *This Fabulous Century*, volume 1, 1900-1910 (New York: Time-Life Books, 1970), 61.

27 "Hadley's Blind Poet," *Daily Hampshire Gazette* 13 February 1906, page 8, columns 5-8, quoting the *New York Sun*. Perhaps his father's mental condition contributed to the careless shooting. In November 1882, Enos had traveled to High Point, North Carolina, to attend a dog show. There he "went crazy" ("Berkshire County," *Springfield Daily Republican* 20 November 1882, page 6, column 4. Also source of "insanity in his family" phrase below). The incident was widely reported. An intriguing notice in the Greenfield paper read: "E. S. Hawkes, who went to North Carolina to attend a field trial of dogs about election time and was reported from southern sources to have gone insane, has returned home, and makes a long statement concerning his experiences in the South. His story will appear in these columns next week" ("Ashfield," *The Gazette and Courier* 11 December 1882). Unfortunately, no follow up story appeared in the paper.

Reporters offered two different reasons for Enos' breakdown: "While there is insanity in his family, Hawks's present condition is attributed to a severe attack of vertigo from which he suffered while visiting a married sister at Florence recently." The only corroborating evidence for this claim is the fact that Enos' paternal grandmother Abigail "Nabby" Marsh (c. 1772-1842) did end her days "at the Worcester Insane Hospital" ("Died," *The Gazette and Courier* 29 November 1842, page 3, column 2. Data from Shirley Majewski, Pocumtuck Valley Memorial Association Library, Deerfield, MA. February 2007).

Another explanation came out eleven days after the first notice of his homecoming: "E. S. Hawks of Ashfield has returned home and appears to be much improved, but still tells some remarkable stories of being pursued and of his dogs being killed, because he was a republican and from the North" ("Berkshire County," *Springfield Daily Republican* 23 Nov 1882, page 6, column 3).

I offer still another possible economic cause for Enos' psychological disorder that is based upon his being a cousin of William Smith Clark (1826-1886). The flamboyant Amherst politician, teacher, traveler to

Germany and Japan, first sitting president of Massachusetts Agricultural College (now the University of Massachusetts), as well as companion and perhaps lover of Emily Dickinson, had urged friends, neighbors and relatives to buy stock in western gold, silver and coal mines. By spring 1882, the scheme had collapsed, ruining many and leaving Clark in disgrace. Although Enos (as the Springfield newspaper account of 20 November 1882 described him) was "quite poor and has a wife and four children," he may have anticipated the lump sum pension, borrowed against it to invest in Clark's scheme and lost all, exacerbating his poverty and his psychological problems.

28   Clarence Hawkes, *Big Brother. The Story of a Trick Bear* (New York: Platt & Munk, 1930), 185-86.

29   Bruce Barton, "God Took My Eyes That My Soul Might See," *American Magazine* 102 (July 1926): 80, 82. Reprinted as "Here is a Remarkable Story That Answers All Those Who Have any Quarrel With Their Circumstances," *Boston Sunday Post* 23 November 1930, Color-Feature Section, page 2. Reprinted in Hawkes, *My Country*, 149-70, along with appreciative letters responding to it.

30   "Serious if Not Fatal Accident at Spruce Corner, Ashfield," *Daily Hampshire Gazette* 21 August 1883, page 2, column 9.

31   The actual fates endured by handicapped men confirmed the dread in fiction. Hawkes' future home, Hadley, had sent 233 men to the Civil War. 17% of them had died; 15% were wounded, and 12% discharged for some disability. Many of these veterans still lived in town when Hawkes arrived in 1892 (Eric N. Freeman, *The March to See the Elephant: Hadley's Participation in the Civil War.* Unpublished junior essay, Deerfield [Massachusetts] Academy, 22 February 1991). A few received tiny awards for service-connected injuries: $4.00 per month for Francis Wheeler's lost thumb, $2.00 per month for Lucius D. Smith's disabled leg (*List of Pensioners on the Roll*, January 1, 1883, 2 volumes [Baltimore: Genealogical Publishing, 1883; reprint 1970], 1. 369). Hawkes several times honored such sacrifices in poems and speeches.

32   "Notable Men " *Daily Hampshire Gazette* 13 October 1908,

page 10, column 5.

33 "In Regard to Helen Keller" *The Critic* 27 (17 April 1897): 792.

34 Review of *Hitting the Dark Trail*. In *The Bookman* 51 (December 1916): supplement 36.

35 William T. Heisler, "An Interview with Perkins' Past" *The Lantern* 47. 2 (March 1978): 12. Hawkes paid tribute to the lessons he learned in "How We Lived at Perkins Institute," *The Outlook* 89 (6 June 1908): 301-305. The purposeful schedule at Perkins resembled earlier ones implied by Thomas More's *Utopia*, established by Ben Franklin, parodied by Gilbert and Sullivan's *Gondoliers*, and adhered to by young Jimmy Gatz, the upwardly mobile hero of F. Scott Fitzgerald's *The Great Gatsby*. From the school's founding, students rose at 5 A.M., assembled at 6 in the chapel, ate breakfast at 8, after which the boys walked and the girls did house work. 9-10 offered school lessons, followed at 11 for community singing. A half hour recess ended so pupils could return to class from 11:30 to 1 P.M. Then dinner and recess, followed from 2-6 with work. 6-7 furnished supper, 7-8 more singing, while 8-9 allowed "reading newspapers, and history" (*Fifth Annual Report of the Trustees of the New-England Institution for the Education of the Blind* [Boston: Press of the Boston Courier, 1837], 7-8). The academic courses mirrored those of other schools with the addition of mechanical aids such as "Eaton's wooden ciphering-boards"–devices with squares marked out by brass strips so students could place numerals in the proper position (*Fifty-Sixth Annual Report of the Trustees of the Perkins Institution and Massachusetts School for the Blind for the Year Ending September 30, 1877* [Boston: Rand, Avery, 1878], 68. A similar wood, leather and metal device that guided Hawkes' handwriting now rests in the Goodwin Memorial Library, Hadley).

36 Barton, "Here is a Remarkable Story," 2.

37 Quoted in Paul M. Kelly, "Adversity as No Obstacle: The Life and Times of Poet and Naturalist, Clarence Hawkes," MA Thesis, Westfield State College, May 1992, page 11.

38 *Hopkins Academy Through the Years ... 1886-1964*, ed. Miriam

R. Pratt, Phyllis E. Podolak (Amherst: Hamilton Newell, 1964), sv 1890; 1894.

39   Grady forthrightly noted that slavery had gone, "But the freedman remains. With him a problem without precedent or parallel." Still, one can now see "a thousand happy Negroes, happy in their cabin homes"; "there are Negro lawyers, teachers, editors, dentists, doctors, preachers"; they may have lost political power, but it will be returned when they are no longer "clannish, credulous, impulsive and passionate." Grady ended with a rhetorical flourish that assumed black citizens could make choices:

> Whatever the future may hold for them—whether they plod along in the servitude from which they have never been lifted since the Cyrenian was laid hold upon by the Roman soldiers and made to bear the cross of the fainting Christ—whether they find homes again in Africa, and thus hasten the prophecy of the psalmist who said, "And suddenly Ethiopia shall hold out her hands unto God"—whether, forever dislocated and separated, they remain a weak people beset by stronger, and exist as the Turk, who lives in the jealousy rather than in the conscience of Europe—or whether in this miraculous Republic they break through the caste of twenty centuries and, belying universal history, reach the full stature of citizenship, and in peace maintain it—we shall give them uttermost justice and abiding friendship.

(Henry W. Grady, "The Race Question." [1889. Reprint. Birmingham: DeBardeleben Coal Corporation, 1956], 1-5)

The same stereotypical optimism appeared years later in Hawkes' *Big Brother*. A circus train winds "its slow way through Virginia and the Carolinas and into the Gulf States. Here the scenes were quite different from those in the north, for they saw the Negroes picking cotton in the field and heard their melodious songs as they sang at their work" (109-110).

40   *Boston Transcript* 4 June 1890.

41   *Boston Journal with Supplement*, Morning Edition 4 June 1890.

Perkins data from Jan Seymour-Ford, Research Librarian, Perkins School, via phone, mail and email, October-December 2006.

42   Hawkes, *Igloo Stories*, 116-17.

43   Helen H. Foster and William W. Streeter, *Only One Cummington. A Book in Two Parts* (Cummington: np, 1974), 313.

44   Hawkes, *Hitting the Dark Trail*, 107.

45   "Ashfield," *The Gazette and Courier* 23 January 1892, page 5, column 1.

46   "Fifty Years of Blindness No Great Handicap to Hadley's Famed Poet," *Holyoke* [Massachusetts] *Daily Transcript* 12 August 1933.

47   "Dr. Hawkes, Blind, Near 80; Poet and Sports Authority," *Springfield Sunday Republican* 11 December 1949.

48   *Hopkins Academy, 1667-1895* (Springfield: Henry R. Johnson, nd), 16.

49   Hawkes, *Hitting the Dark Trail*, 125.

50   Hawkes, *Notes of A Naturalist*, 57-58.

51   1869 Birth Records, Hadley Town Hall. 1899 Marriage Certificate, Hadley Town Hall. Hawkes specially thanks "My brave little mother-in-law" who "did much of my reading." (*Hitting the Dark Trail*, 128)

52   Pratt data from Paul Schlotthauer, Archivist, Pratt Institute. November 2006.

53   Built about 1760, the large white structure housed the Smith family until 1870. Then Bessie's father Samuel bought it from his aunts, Ruth (wife of Ephraim Smith) and Maria (wife of Elijah Smith), who had lived there since their father-in-law Seth died in 1828. This house, Hawkes' home, and the double row of trees that march resolutely north and south along the beautiful common still stand, witnesses to New England tradition.

54   Hawkes, *Hitting the Dark Trail*, 116.

55   "The Blind Poet Married," *Daily Hampshire Gazette* 30 October 1899, page 7, column 2.

56   "Clarence Hawkes is 66 Years Old Today," *Daily Hampshire*

Gazette 16 December 1935, page 1, column 5.

57 Hawkes, *Notes of A Naturalist*, 18.

58 Hawkes, *The Service Man's Friend* (Boston: Chapman & Grimes, 1944), 28.

59 "Clarence Hawkes Speaks at Goshen," *Daily Hampshire Gazette* 15 August 1931, page 2, column 6.

60 Hawkes, *The Service Man's Friend*, 51. Howe may have introduced Hawkes to the Vice President of the Authors' Club, Thomas Wentworth Higginson (1823-1911). Hawkes' restatement of Abolitionist beliefs in his graduation speech would have been congenial to Higginson, who had led black soldiers during the Civil War, and Hawkes' verses might have impressed him—he co-edited Emily Dickinson's poems. Or his cousin, Henry Lee Higginson, a Trustee and member of Perkins' corporation, may have noticed the young writer.

Other members were also familiar with him. Mable Loomis Todd (1856-1932), friend of Higginson and the other editor of Dickinson, plus Mable's husband, David Peck Todd (1855-1939), an astronomy professor at Amherst College, knew him. The year before, she had recognized his talents and wondered in an appreciative letter whether he would like to speak under the auspices of the Amherst Historical Society before an audience of "fifty or sixty," a welcome change from rustic buildings and few auditors (Letter from Mable Loomis Todd to Hawkes, 27 December 1899, Hawkes Manuscript Letters 1. 40).

The Boston group contained authors whose interests coincided with Hawkes'. Arthur Gilman (1837-1909) distinguished himself as a genealogist, as a historian of Rome and the American people, and as editor of Geoffrey Chaucer. His plan to provide a college education for women may have influenced Hawkes' later campaign to form female conservation groups. The patriotic words of Katharine Lee Bates' (1869-1929) "America, the Beautiful" consoled him in Hadley during the Second World War (*The Service Man's Friend,* 50). Finally, John Townsend Trowbridge (1827-1916) wrote books for boys that stressed adventure and clean living, two staples of Hawkes' own works.

61   Hawkes, "A Night with Ruff Grouse," *Woman's Home Companion* 31 (November 1904): 49.

62   "Fifty Years of Blindness No Great Handicap to Hadley's Famed Poet." *Holyoke Daily Transcript* 12 August 1933.

63   Copy of letter from Frederick G. Howes to Hawkes, 3 December 1901. Ashfield Historical Society. Accession number 4307e.

64   (Salem, MA: Samuel Edson Cassino & Son; New York & Boston: H.M. Caldwell, 1904). Four other anthologies demonstrate the authors judged to be Hawkes' companions. *Junior Literature. Seventh Year*, ed. Walter L. Hervey (New York: Longmans, Green, 1929) juxtaposed "The Thundering Herd" and "Pep: A Blue-Ribbon Dog" with works by Hamlin Garland, Vachel Lindsay, Bret Hart, William Wordsworth, Washington Irving, Hugh Lofting, Robert Browning, Sir Walter Scott, and Rudyard Kipling. Another school text, *Elson Junior Literature. Book One*, ed. William H. Elson, Christine M. Keck, Mary H. Burris (Chicago, etc.: Scott, Foresman, 1932), reprinted "A Night with Ruff Grouse" next to writings by Henry David Thoreau, Edwin Markham, Ralph Waldo Emerson, William Shakespeare, Nathaniel Hawthorne, and Mark Twain. Yet a third gathering, *Elson Junior Literature. Book Two*. Editors William H. Elson, Christine M. Keck, Mary H. Burris (Chicago, etc.: Scott, Foresman, 1932), offered "The Thundering Herd" alongside essays by Ernest Thompson Seton, Theodore Roosevelt, William Cullen Bryant, Sidney Lanier, John James Audubon, Dallas Lore Sharp, Thomas Hardy, and Henry W. Grady. A fourth collection, *Earth and Sky. Book Four,* ed. Paul Grey Edwards and James Woodward Sherman (Boston: Little, Brown, 1937), began with Hawkes' "A Cradle in the Treetop," followed by works by Emily Dickinson, John Keats, Charles and Mary Lamb, Joseph Addison, William Blake, and Amy Lowell.

65   Letter from Hugh Lofting, 26 December 1923, Hawkes Manuscript Letters 1. 173.

66   Edgar Rice Burroughs' material available at *www.erbzine.com*

67   Howard Clark Brown, "A Survey of the Naturalistic Periodical

Literature of America," *American Midland Naturalist* 7. 3 (May 1921): 75-76.

68 Grenville Goodwin, "Excerpts from the Apache Diaries," *Journal of the Southwest* (Spring-Summer 2000): 4.

69 "With The Overland Monthly Contributors," *The Overland Monthly* 46 (October 1905): 366.

70 Advertisement for Henry Holt & Company, *The New York Times* 9 October 1915, page 12.

71 "Master's Degree Conferred upon Clarence Hawkes," *Daily Hampshire Gazette* 14 June 1917, page 2, column 3. In May 2006 Linda Benedict of the Hobart-William Smith Library kindly furnished a commencement program that lists the 21 graduates, only two of whom had majored in English. Latin, French, and History were more popular.

72 The Syracuse archives contain Hawkes' 1947 request for a duplicate certificate: the 1942 fire at Northampton's Forbes Library had destroyed the original plus "six honorary diplomas which I had seen f [sic] received from different American Colleges" (Typescript letter from Hawkes to Chancellor Graham of Syracuse, 22 September 1947. Data from Mary O'Brien, Archives and Records Management, Syracuse University Library, October 2006).

73 "Amherst Alumni Dine," *The New York Times* 19 June 1919, page 11.

74 George F. Whicher, "The Victory Commencement," *Amherst Graduates' Quarterly* 8. 4 (August 1919): 122. The connection with Amherst lasted. President Stanley King congratulated Hawkes after United Press commemorated his sixty years of blindness in 1400 newspapers:

> No one takes a warmer pride in your accomplishment than the College, which is proud to count you as her foster son. Each generation of college students at Amherst has known you, and the young men who have played on our athletic teams have thought of you as perhaps their most consistent friend.

(Letter from Stanley King to Hawkes, 8 September 1943. Amherst

College Archives.)

75 "Principals at A. I. C. Commencement," *AIC Yellowjacket* 8 June 1938. Furnished by Katherine DeLiso, Shea Memorial Library, A. I. C. December 2006.

Just as Hawkes retained data from (as Bartlett's A. I. C. citation put it) "maps, atlases, geographies and natural histories," so he held on to friends. The A. I. C. Commencement Program noted that Professor Dallas Lore Sharp, Junior (1900-1970), awarded the BA degrees. Three decades before the ceremony, his father, Dallas Lore Sharp, Senior (1870-1929), the Boston naturalist who, like Hawkes, provided a powerful voice for the ideals of conservation and preservation, had begun a fond correspondence with Hawkes. Sharp's first letter set the tone for their mutually respectful relationship: "Every genuine nature-lover appreciates your work, and every lover of courage, patience and the gentleness that makes men great sympathizes with you and admires you profoundly" (Letter from Dallas Lore Sharp, 6 May 1906. Hawkes Manuscript Letters 1. 63). The two reconfirmed their friendship when the Sharps visited the Hawkes at Hadley, a trip that in 1915 required nine hours each way from Boston but, because of the amity, "would have made twice the journey infinitely worth taking" (Letter from Dallas Lore Sharp to Hawkes, 5 July 1915. Hawkes Manuscript Letters, 1. 107). On another visit that fall, Sharp jocularly promised to bring his sons (probably including Junior), "If you can give us boys bunks on the floor" (Letter from Dallas Lore Sharp, 29 September 1915. Hawkes Manuscript Letters, 1. 112).

76 Percy F. Bicknell, "Fawcett and Hawkes, Parallels of Tragedy," *The Baltimore Sun*, 19 July 1923, page 5. Reprinted in Hawkes, *My Country*, 130-42.

77 "Clarence Hawkes." *Nature Magazine* 37. 2 (February 1944): 104. Hawkes fondly refers to "one of his [Hornaday's] famous books" in *Big Brother*, 90. Hornaday, in turn, wrote the admiring preface to Hawkes' *Trails to Wood and Water*.

78 Hawkes, *The Light That Did Not Fail*, 32.

79 "Clarence Hawkes Enjoys College Baseball," *Daily Hampshire Gazette* 21 June 1916, page 5, column 5.

80 "Clarence Hawkes Attends Ball Game in New York." *Daily Hampshire Gazette* 15 July 1917, page 2, column 3.

81 "Clarence Hawkes Plans for Inter-City Baseball," *Daily Hampshire Gazette* 1 August 1918, page 6, column 2.

82 "Clarence Hawkes Tells How He 'Sees' Contests on the Football Field," *Daily Hampshire Gazette* 30 November 1937, page 7, columns 2-4.

83 "Clarence Hawkes and his Work Described by Writer in *The Worcester Telegram*," *Daily Hampshire Gazette* 11 August 1936, page 4, columns 1-4. Originally in *The Worcester* [Massachusetts] *Telegram*.

84 Undated photograph from unknown newspaper. c. 1936. Ashfield Historical Society. Accession number 1504.

85 Hawkes, *Hitting the Dark Trail*, 164-67.

86 "Hadley, Mass., Aug. 4," *The New York Times* 5 August 1909, page 7, article 11.

87 "'The Man of Marion' By Hadley Author," *Daily Hampshire Gazette* 29 August 1923, page 2, columns 3-4.

88 "Granite And Gold, In Tribute To Coolidge," *Daily Hampshire Gazette* 30 October 1924, page 4, column 3.

89 "Clarence Hawkes on Coolidge," *Daily Hampshire Gazette* 29 April 1933, page 2, columns 3-4.

90 "Clarence Hawkes Writes On Hoover," *Daily Hampshire Gazette* 28 October 1932, page 4, columns 4-6.

91 Hawkes, *Notes of A Naturalist*, 63-68. One example of the nostalgia for times gone by, Charles Clark Munn's 1907 *Boyhood Days on a Farm*, provided the bass for Hawkes' melodies. Munn sadly acknowledged that young people were rapidly deserting New England farms and missing both the glories of nature and of rural discipline. Like Isaiah 34, which gloomily envisioned creation reversing itself, undoing the heavens and leaving only wild beasts amid the ruins of earth's cities, Munn first pictured the scenes of yesteryear ("The old gambrel-roofed

farmhouse with open fireplace, ... the spacious barns that in summer he helped fill with golden grain and fragrant hay, ... the wide meadows, the extended forests, the brimming brooks"), then looked at them today ("the farmhouses falling into decay, or occupied by foreigners, the woods cut away, the brooks dwindling to rills"). The loss of rural place and occupation prevents future generations from developing "a vigorous body and strong mind," qualities central to Jeffersonian democracy. (Charles Clark Munn, *Boyhood Days on a Farm, A Story for Young and Old Boys* [Boston: Lothrop, Lee & Shepherd, 1907], ii-iv).

    92   Hawkes, *Hitting the Dark Trail*, 19.

    93   Henry David Thoreau, *A Week on the Concord and Merrimac Rivers*, ed. Carl F. Houde, William L. Howarth, Elizabeth Hall Witherall (1849. Princeton: PUP, 1980), 34.

    94   William Wordsworth, "The Tables Turned," in *The Poetical Works*, ed. Thomas Hutchinson and Aubrey de Selincourt (London: Oxford UP, 1960), 377.

    95   William Bradford, "Of Plymouth Plantation" in *The American Tradition in Literature*, Third Edition, ed. Sculley Bradley, Richmond Croom Beatty, E. Hudson Long (New York: W. W. Norton, 1967), 1. 19.

    96   Perry Miller, "Nature and the National Ego," in *Errand into the Wilderness*, (Cambridge, MA: Belknap Press of Harvard UP, 1956), 205-16.

    97   Flood data from *Souvenir Booklet, Hadley Tercentenary, 1659-1958, July 31, August 1, 2*, ed. Mrs. John T, Martula, Richard Martula, Susan Martula. Hawkes was spared the rumor spread by "two men under the influence" who paddled a canoe to any house on West Street where people still remained and warned them a 60' wall of water was approaching. "People who recognized the tipplers invited them to drown first."

    98   Hawkes, *Notes of A Naturalist*, 79.

    99   Hawkes, *My Country*, 104.

    100 Ibid., 200.

    101 "Clarence Hawkes Mourns His Pet," *Daily Hampshire Gazette*

30 April 1930, page 4, columns 5-7. Also in Springfield, Greenfield and unattributed newspapers.

102 "A Tribute To Animals," *Daily Hampshire Gazette* 19 April 1933, page 4, columns 3-4.

103 "Hadley Poet's Only Anxiety Is to Be Worthy of His Dog," *The Springfield Union* 3 December 1931.

104 Dallas Lore Sharp, "Introduction," in Hawkes, *Hitting the Dark Trail*, xi.

105 Hawkes, *The Light That Did Not Fail*, 130-31.

106 Alexander Pope, "An Essay on Criticism" 1. 298, in *The Poems*, ed. John Butt (New Haven: Yale UP, 1963), 153.

107 Kenneth Burke, *The Philosophy of Literary Form* (New York: Vintage, 1957), 353.

108 Hawkes, *Stories of the Good Green Wood* (New York: Thomas Y. Crowell, 1904), 48.

Also quoted in *Notes of a Naturalist*. 13.

109 Hawkes, "Ma's Posy Beds," in *Idyls of Old New England* (Northampton: Picturesque Publishing, 1897), 57.

110 Hawkes, "How Massa Linkum Came," in *Pebbles and Shells* (Northampton: Picturesque Publishing, 1895), 79. Lincoln's son Robert Todd Lincoln wrote Hawkes an appreciative note on 28 June 1895 that also kindly added, "Permit me also to congratulate you at the bravery and cheerfulness with which you bear your heavy burden" (Hawkes, *The Light That Did Not Fail*, 108). Hawkes tells how he learned the facts for the poem in *My Country. The America I Knew*, 54-63. Several radio stations broadcast the poem.

111 Clarence Hawkes, *Three Little Folks. Verses for Children* (Northampton: Picturesque Publishing, 1896), 13-14.

112 Review of *Three Little Folks* appended to Clarence Hawkes, *Songs for Columbia's Heroes* (Springfield: New England Publishing, 1898), 20.

113 Clarence Hawkes, *The Hope of the World* (Springfield: New England Publishing, 1900), 29-30.

114 Hawkes, "Human Flowers," in *Idyls of Old New England*, 29.

115 Hawkes, "Keep Up Yer Fences," in *Idyls of Old New England*, 99.

116 Hawkes, "The Fate of the Maine," in *Songs for Columbia's Heroes*, 37.

117 Hawkes, "Remember the Maine," in *Songs for Columbia's Heroes*, 44.

118 Hawkes, "The Coming of Peace," in *Songs for Columbia's Heroes*, 72.

119 Hawkes, "Columbia and Britannia," in *Songs for Columbia's Heroes*, 73.

120 Hawkes, "The Ole Meetin-House," in *Idyls of Old New England*, 76.

121 Clara Clough Lenroot, *Long, Long Ago* (Appleton, WI: Badger Printing, 1929), 30.

122 Hawkes, *Hitting the Dark Trail*, 56.

123 "Afield with a Blind Naturalist," *The Outing Magazine* 51 (December 1907): 347.

124 Hawkes, *Igloo Stories*, 96, 99.

125 Quoted in Hawkes, *The Light That Did Not Fail*, 123. Also in Hawkes, *My Country*, 86.

126 Hawkes, *The Service Man's Friend*, 48.

127 Hawkes, *King of the Flying Sledge*, 10.

128 Clarence Hawkes, *Shaggycoat; the Biography of a Beaver* (Philadelphia, G. W. Jacobs, 1906), passim.

129 Mark Twain, *Adventures of Huckleberry Finn*. Second edition, ed. Sculley Bradley, Richmond Croom Beatty, E. Hudson Long (1885. New York: W. W. Norton, 1977), 229.

130 "Birding by Ear," *Audubon* 50 (November 1948): 352.

131 Letter from Thornton Burgess, 25 February 1930. Hawkes Manuscript Letters 1. 199.

132 Clarence Hawkes, "Old Ringtail's Waterloo" *Outing Magazine* 45. 2 (November 1904): 201-205.

133 Originally Jack London, "Diable – A Dog," *The Cosmopolitan* 33 (June 1902): 218-26. Finally, "Bâtard." *The Faith of Men and Other Stories* (New York: Macmillan, April 1904).

134 Clarence Hawkes, *Silversheene, King of Sled Dogs* (Springfield, MA: Milton Bradley, 1924), 97.

135 Clarence Hawkes, *Pep. The Story of a Brave Dog* (Springfield, MA: Milton Bradley, 1922), 91.

136 Ernest Hemingway, "A Way You'll Never Be," in *The Complete Short Stories* (1932. New York: Charles Scribner's Sons, 1987), 306.

137 Lawrence Buell, *The Environmental Imagination: Thoreau, Nature Writing, and the Formation of American Culture* (Cambridge, MA: Harvard UP, 1995), 7-8.

138 Max Oelschlaeger, *The Idea of Wilderness From Prehistory to the Age of Ecology* (New Haven: Yale UP, 1991), 294. Hawkes' accuracy saved him from blame during the decades long "nature faker" controversy. In 1903, John Burroughs' "Real and Sham Natural History" (*The Atlantic Monthly* 91.545 [March 1903]: 298-309) derided those authors who purported to describe the animated world but offered a gullible public only sentimental and inaccurate ideas of the non-humun world. Ralph H. Lutts' *The Nature Fakers* (Charlottesville and London: University of Virginias Press, 1990) traces the change in attitude about human relations to animals. President Roosevelt supported Burroughs and also replied to Hawkes' correction of his misstatement about moose.

139 Saint Augustine, The Confessions 8. 12.

140 Hawkes, *Wanted a Mother*, 25. The cold language and three characters may owe something to two popular works that Hawkes probably knew. In 1864 Elizabeth Barrett Browning told of another unloved charge. Aurora Leigh's parents died in Italy so she must return to England as the ward of her father's hostile sister. Although the aunt had loved her brother, she hated the Tuscan mother and thus dislikes Aurora.

She stood upon the steps to welcome me, Calm, in black garb.

> I clung about her neck,
> Young babes, who catch at every shred of wool
> To draw the new light closer, catch and cling
> Less blindly. In my ears, my father's word
> Hummed ignorantly, as the sea in shells,
> 'Love, love, my child,' She, black there with my grief,
> Might feel my love–she was his sister once–
> I clung to her. A moment, she seemed moved.
> Kissed me with cold lips, suffered me to cling,
> And drew me feebly through the hall, into
> The room she sate in.
> There, with some strange spasm
> Of pain and passion, she wrung loose my hands
> Imperiously, and held me at arm's length,
> And with two grey-steel naked-bladed eyes
> Searched through my face,–ay, stabbed it through and through,
> Through brows and cheeks and chin, as if to find
> A wicked murderer in my innocent face,
> If not here, there perhaps.

(Elizabeth Barrett Browning, *Aurora Leigh* [London: J. Miller, 1864], Book One)

Aurora goes on to become a writer, perhaps predicting what Eleanor will do if she follows the life plan Nathan has offered, that of a college graduate and author.

The triangle of harsh mistress, weak but benign male and pretty interloper who enchants the man sounds like Edith Wharton's grim *Ethan Frome* (1911), also set on a farm in the hills of western Massachusetts. Like Eleanor, Maddie, the spirited but poor relation who must board with her unwilling aunt, breaks one of the wife's prized glass dishes; in Hawkes, the child breaks both a dish and window. Maddie can orate; Eleanor "had learned a piece of poetry at the poor-farm ... [and] had recited it with great success before the inmates of the farm at a Christmas entertainment" (133). The parallels end because Hawkes rejects Wharton's fatalism—her wife never accepts Maddie, rightly suspecting that her husband has fallen in love with her, so the pair try unsuccessfully to commit suicide and merely cripple themselves.

141 Robert Browning, "Pippa Passes. A Drama," in *Bells and Pomegranates* (1841). Quoted in Hawkes, *The Service Man's Friend*, 19.

142 The sunshine phrase now seems mawkish, but in its day American ears probably accepted it without criticism. The popular radio program *Vic and Sade*, which from 1932-1947 chronicled with affection and humor the trivial events of small town life in Illinois, once had husband Vic return home in a "jolly" mood. He explains to his wife that a reporter from their local newspaper asked about his trip "down South" to Kentucky, Indiana and Ohio, so Vic handed in 15 pages of description. "I feel good," he boasts, "because my soul is a buttercup which has caught the liquid sunshine in its golden chalice" ("Vic's Geographical Trip." 1939).

143 Frederick G. Howes, *History of the Town of Ashfield Franklin Count, Massachusetts from its Settlement in 1742 to 1910* (Ashfield: Town of Ashfield, 1910), 334.

144 *University Library of Autobiography Including All the Great Autobiographies and the Autobiographical Data Left by the World's Famous Men and Women*, Volume 15 (np: National Alumni, 1927).

145 "A Son of Goshen," *Daily Hampshire Gazette* 9 July 1931, page 8, column 2.

146 "Clarence Hawkes Honored," *Daily Hampshire Gazette* 25 July 1936, page 8, column 7.

147 "Clarence Hawkes Given Three Degrees by the Knights of Pythias," *Daily Hampshire Gazette* 14 December 1937, page 6, column 3.

148 "Clarence Hawkes Made an Honorary Member of the Northampton Lions Club," *Daily Hampshire Gazette*. Undated clipping.

149 Robert Bartlett, *They Dared to Live* (New York: YMCA, 1937).

150 Bunny Stein, "The Roaring Twenties," *http://www.angelfire.com/va2/twigs/pages/pages/r20s*

151 Bessie occasionally drove. The couple owned a 1932 Ford and then, moving slightly upward in status, a 1937 Chevrolet. The kind of vehicle "demolished" when hit on 14 October 1932 at the corner of Northampton and Dartmouth streets in Northampton was not specified, but it careered through the intersection and turned over, requiring medical people to transport the shocked and bruised Hawkes couple to Holyoke Hospital for treatment (*Daily Hampshire Gazette* 15 October 1932, page 1, column 5).

152 McQueston personal interview, Monday, 23 October 2006.

153 John Riley, "Seeing the World Along with Hadley's 'Blind Poet,'" *Amherst* [Massachusetts] *Bulletin* 4 September 1992, page 11.

154 Woicekoski personal interview, Wednesday, 22 November 2006.

155 Letter from Hawkes, *Daily Hampshire Gazette* 1 October 1915, page 4, columns 4-5.

156 Illinois Civil War Records, *www.ilsos.gov/genealogy/CivilWarController*

157 Niedbala personal interview, Tuesday, 12 December 2006.

158 One helper, Alessandro Capabianca (1867–1928), was nearly the same age as his host. Nicknamed "Sandy," according to McQueston, the Hawkes memorialized him in two ways. First, they paid for his small but dignified marker that now lies next to theirs in the Old Hadley Cemetery. More personally, Hawkes published "Angela–A Love Story" during 1917. In it, a winsome immigrant named Angela peddles trinkets to earn money so her sweetheart can travel from Italy to America. His name is unItalian but significant: Alesandra.

159 Jakobek personal interview, Tuesday, 17 October 2006.

160 Pratt telephone interview, Monday, 23 October 2006.

161 Lesko personal interview, Saturday, 11 November 2006.

162 Words and music for the first song appear in Margaret Clifford

Dwyer, *Hopkins Academy and the Hopkins Fund, 1664-1964* (Hadley: Trustees of Hopkins Academy, 1964), 182-183.

163 "Six Decades in Darkness," *Daily Hampshire Gazette* 14 August 1943, page 6, column 1.

164 Hawkes, *Hitting the Dark Trail*, 85.

165 Ralph Waldo Emerson, "Nature," in *The American Tradition in Literature*, volume 1, third edition, ed. Sculley Bradley and Richmond Croom Beatty (New York: W. W. Norton, 1967), 1066-67.

166 Hawkes, *Notes of a Naturalist*, 8; 11.

167 William Cullen Bryant, "Thanatopsis," in *The American Tradition in Literature*, 473-74.

168 "Notable Books in Brief Review," *The New York Times* 25 November 1917, page 501.

169 Alfred Lord Tennyson, "In Memoriam" 56, in *The Poems*, ed. Christopher Ricks; (London: Longmans, 1969), 912.

Cover photograph. Clarence Hawkes at the North Hadley Pond. Undated. (1936?) Courtesy of the Goodwin Memorial Library, Hadley

Genealogy House
Publishers of Family History and Genealogy

Genealogy House publishes narrative family histories and genealogies that combine good writing and editing with genealogical research.

By incorporating high editorial and production standards, we maintain the best of traditional publishing and combine it with today's advantages of digital printing and distribution technologies. The result is a process that is professional, efficient, cost effective, flexible, and responsible.

Our goal is to capture the spirit of the people of the past to share with generations to come.

Genealogy House
a division of White River Press
Amherst, Massachusetts

www.genealogyhouse.net

www.ingramcontent.com/pod-product-compliance
Lightning Source LLC
Chambersburg PA
CBHW032126090426
42743CB00007B/484